IMAGES
of America

THE SCOPES
MONKEY TRIAL

The Scopes Monkey Trial took place in July 1925 in the Rhea County Courthouse in Dayton, Tennessee. The courthouse was declared a National Historic Landmark in 1977, and the large, second-floor courtroom in which Scopes was tried remains operational. Each year, hundreds of people come to Dayton to see the courthouse and learn about Scopes's famous trial, which has become a symbolic landmark in the ongoing clash between science and religion. (Randy Moore.)

ON THE COVER: The legendary Scopes Trial took place in a hot, standing-room-only courtroom in July 1925 in tiny Dayton, Tennessee. In this courtroom scene, defense attorneys Clarence Darrow (left, leaning back against the table) and Dudley Field Malone (standing with coat on behind Darrow) listen to prosecutors during the trial. The youthful defendant, John Scopes, can be seen leaning forward just behind Darrow. (Library of Congress.)

IMAGES
of America

THE SCOPES
MONKEY TRIAL

Randy Moore and William F. McComas
Foreword by Tom Davis

ARCADIA
PUBLISHING

Published by Arcadia Publishing
Charleston, South Carolina

Library of Congress Control Number: 2015955170

For all general information, please contact Arcadia Publishing:
Telephone 843-853-2070
Fax 843-853-0044
E-mail sales@arcadiapublishing.com
For customer service and orders:
Toll-Free 1-888-313-2665

Visit us on the Internet at www.arcadiapublishing.com

*We dedicate this book to John T. Scopes and the many science
teachers who have followed his courageous example.*

CONTENTS

FOREWORD

During the hot summer of 1925, the attention of the world was focused on Dayton, a little town in the hills of Tennessee. Before this time, virtually no one had heard of Dayton, but the Scopes Monkey Trial changed that.

John Scopes, the trial's young defendant, was a coach and science teacher at Rhea Central High School. Just a few blocks away from Scopes's classroom was Robinson's Drug Store, where the chairman of the county board of education presided over an emporium that sold sodas, prescriptions, and schoolbooks. While sitting around a tiny table in the local drugstore, F.E. Robinson, George Rappleyea, Sue K. Hicks, Walter White, and other city leaders laid plans for what became a landmark in American history.

The old high school and drugstore are gone, and Dayton has grown significantly since it hosted Scopes's famous trial. But the Rhea County Courthouse still stands, shaded today by a number of trees that provided respite from the blazing sun that beat down on the thousands who gathered to see the legal fireworks. These people included two of the most famous lawyers in America: William Jennings Bryan, one of the country's leading spokesmen for fundamentalist Christianity and a longtime leader of the Democratic party, and Clarence Darrow, arguably the premier defense attorney of his day, a noted agnostic, and former political supporter of Bryan. With them were local officials such as District Attorney Tom Stewart and attorney Ben McKenzie and the often overlooked defendant, John Scopes.

Much remains of the charm and appearance of 1925 Dayton, but one thing has changed: the Scopes Trial certainly put Dayton on the map. Hardly a day passes without someone stopping in town to see where the trial was held. The hometowns of our visitors are spotted all over the world.

We hope Images of America: *The Scopes Monkey Trial* will help you understand John Scopes's famous trial and give you a better appreciation for the people, events, and community that produced "The World's Most Famous Court Trial."

—Tom Davis, President
Rhea County Historical and Genealogical Society

ACKNOWLEDGMENTS

We are indebted to many people who helped us with this project. We found several images in this book with the help of Ryan Greenwood of the Riesenfeld Rare Books Research Center at the University of Minnesota Law Library; Joan Brasher and Donna Pritchett at Vanderbilt University; Jackie Cameron, Julie Stoner, and Jonathan Eaker in the Prints and Photographs Division at Library of Congress; Mary Markey at Smithsonian Institution Archives; Tennessee State Library and Archives; Kris Bronstad at the University of Tennessee-Knoxville; Kristen Kennedy at the American Civil Liberties Union (ACLU); and Steve Smartt, Colleen and Don Fehn, Jennifer Sprague, and Layton Brueske at the First Baptist Church of Minneapolis. We also thank Jimmy and Bobbie McKenzie for meeting with us and providing information about Ben McKenzie and the trial, and administrators at Bryan College for giving us access to its extensive collection of trial materials. We also appreciate the direct and indirect support of the University of Minnesota and the University of Arkansas. We are particularly grateful for funding from the University of Arkansas Parks Endowed Professorship in Science Education. Of course, all commentary, opinions, and errors are ours alone.

We are especially grateful to our many friends in Dayton, Tennessee, including Dean Wilson, Donna Reed Taylor, the late Eloise "Weezie" Purser Reed, and librarian Connie Sanders and the late Richard Cornelius of Bryan College. We also thank Tom Davis and Pat Hawkins Guffey of the Rhea County (Tennessee) Historical and Genealogical Society, for graciously providing photographs, information, guidance, and hospitality during our many trips to Dayton to work on this book. We also were given access to important documents and images by Jerry Tompkins, Jon Epperson, and Susan Epperson. We could not have produced this book without their generous and ongoing help.

Images in this volume appear courtesy of the *New York World-Telegraph & Sun* Collection at the Library of Congress (LC), Bryan College (BC), Rhea County Historical and Genealogical Society (RCHGS), Smithsonian Institution Archives (SI), Tennessee State Library and Archives (TSLA), First Baptist Church of Minneapolis, Minnesota (FBC), and the University of Tennessee-Knoxville's W.C. Robinson Collection of Scopes Trial Photographs, MS-1091 and MS-1018 in MPA-136 (UTK). These sources are referenced throughout the book by the use of the abbreviations listed above.

INTRODUCTION

It was not a trial in the ordinary sense of the word . . . The crowd came out to watch the trial just as people go to watch a football game, or a prizefight, or any other spectacle. The whole community was stirred, although many did not quite understand what it was all about.

—Warner B. Ragsdale,
an Associated Press reporter who covered the Scopes Trial

The Scopes Monkey Trial—a case formally known as *State of Tennessee v. John Thomas Scopes* (Trial Case No. 5232)—was America's first "trial of the century." Although Scopes's trial occurred in tiny Dayton, Tennessee, and involved a low-level misdemeanor, it became a famous event, cultural landmark, and a topic of legend. The trial, and the issues surrounding it—started for economic reasons and fostered for political ones—quickly captivated the world and inspired countless analyses, documentaries, books, speeches, and sermons. Never before or since has Dayton received such publicity.

How Did the Scopes Trial Originate?

In 1859, British naturalist Charles Darwin published his influential book *On the Origin of Species*, in which he proposes his theory of evolution by natural selection. Although Darwin's book was criticized by British clergy, it was relatively uncontroversial in the United States, where people were preoccupied with events that were inexorably leading to the Civil War. Americans were also comforted by scientists who reconciled evolution and their Christian faith. One such scientist was Harvard biologist Asa Gray, a prominent evangelical Christian who defended Darwin's ideas while reassuring people that evolution was directed by God and therefore not a threat to their faith.

By the end of World War I, however, religious attitudes in the United States had shifted. A collective nostalgia for the relative simplicity of prewar life, combined with a perceived decline in the nation's morality, led many to increasingly look to and rely on their religious faith for stability and comfort. Religious fundamentalism, based on a literal interpretation of the Bible, became popular.

Leading the fundamentalists was Minneapolis preacher William Bell Riley, who in 1919 founded the World Christian Fundamentals Association (WCFA) to organize like-minded believers to resist these changes in society. When more than 5,000 people showed up for Riley's initial meeting of the WCFA, Riley declared the meeting as "an event of more historic moment that the nailing up, at Wittenberg, of Martin Luther's ninety-five theses." As we will see, Riley's organization, the first major nondenominational organization of fundamentalists, would later play a critical role in producing the Scopes Trial.

Riley focused his organization on what he denounced as the "unscientific, anti-Christian, atheistic, anarchistic, pagan" evils of evolution. As the fundamentalist movement gathered steam, countless other preachers spread Riley's antievolution campaign across the country. In the Northeast, militant fundamentalist John Roach Straton proclaimed, "The great battle of the age is now on between Christianity and evolution . . . it is better to wipe out all the schools than undermine the Bible by permitting the teaching of evolution." In the South, fundamentalist firebrand John Norris had evolution-endorsing professors fired at Baylor University and Southern Methodist University, all the while likening evolution to "the poisonous gas of German armies . . . sweeping through our schools." In the West, Aimee Semple McPherson warned her giant congregation that evolution was "poisoning the minds of children" and was responsible for "jazz, crime, and student suicides." Elsewhere, other evangelists blamed evolution for the four "p's"

—perversion, prostitution, permissiveness, and pornography. The sermons delivered by these and other fundamentalist preachers had a tremendous impact. By early 1923, Riley knew that "the whole country is seething on the evolution question."

Riley and his followers, who believed that the teaching of evolution was the intellectual source of moral disorder, wanted a law banning Darwin's ideas from public schools. They got what they wanted when, in March 1925, Tennessee passed a law drafted by legislator John Washington Butler making it "unlawful for any teacher in . . . [Tennessee's] public schools to teach any theory that denies the story of the Divine Creation of man as taught in the Bible, and to teach instead that man is descended from a lower order of animals." Fundamentalists may have viewed such a law a return to prewar normalcy, but famed attorney Clarence Darrow attributed it to the "brainless prejudice of soulless religio-maniacs." In Dayton, Tennessee—a sleepy, religious town 30 miles north of Chattanooga—businessmen hoping to use a sensational trial to help the area's struggling economy found a first-year schoolteacher named John Scopes who was willing to be arrested for violating the law and stand trial. The result was the legendary Scopes Trial.

The Scopes Trial and Its Legend

In Images of America: *The Scopes Monkey Trial*, you will experience a trial that was labeled the "trial of the century" before it even started. Scopes's famous trial included numerous unusual events, as well as a variety of "firsts," such as:

• The Scopes Trial, one of the top news stories of the 20th century, was the first trial in the United States to be broadcast live on radio nationwide. There are no recordings of the trial, however, because technology for recording live radio broadcasts had not been developed.

• Despite its fame, the Scopes Trial produced no legal precedents.

• Although Scopes's trial later attracted the attention of the world, it began as publicity stunt concocted in a local drugstore to improve the area's failing economy. For example, to attract attention, trial instigator George Rappleyea told reporters that Scopes would be defended by famed British writer—and author of *War of the Worlds*—Herbert George "H.G." Wells. When Wells heard of Rappleyea's claim, he rejected it, adding, "I have never heard of Dayton."

• John Scopes, the defendant, was a popular coach and teacher who was not even at school on the day listed in his indictment (April 24). He was not sure he had taught evolution, and if he had, he had done so "without thought that I was violating the law." Regardless, he quickly became irrelevant as the focus of the trial shifted to fundamentalism versus modernism, science versus religion, and the "new" versus the "old."

• Virtually no one in Dayton had an intellectual interest in testing the newly passed law. As Scopes later noted, the trial was simply "a drugstore discussion that got past control." Scopes agreed to participate in the trial not to defend Charles Darwin or evolution (of which he knew relatively little), but instead to defend the overarching cause of academic freedom.

• Helping prosecute Scopes was populist politician and former congressman, secretary of state, and presidential candidate William Jennings Bryan, who hoped to use the evolution controversy to return to national relevance. Bryan, who had championed women's suffrage and a progressive income-tax, was well-respected and popular, thanks largely to his speeches and his syndicated column that appeared in more than 100 newspapers and was read by more than 13 million people. To Bryan, who believed that the teaching of evolution would "take religion out of the hearts of the children," the issue in Tennessee was simple: "Parents have a right to say that no teacher paid by their money shall rob their children of faith in God" by teaching evolution. Bryan promised fundamentalists that their campaign to ban the teaching of evolution would "sweep the country" and that the Scopes Trial was merely the first step in the upcoming war to restore America's morality. In private, Bryan accepted the evolution of "lower" plants and animals, but not of humans. Before coming to Dayton, Bryan offered to pay Scopes's fine.

• Helping defend Scopes was famed criminal defense attorney Clarence Darrow, who volunteered for the case the day after Bryan announced his involvement in the trial. This was the first and only time that Darrow—a onetime political ally of Bryan—volunteered his services "without fees or expenses, to help the defense" of a client. Darrow, who argued that Scopes had been "indicted for the crime of teaching the truth," attributed Tennessee's prosecution of Scopes to "ignorance and fanaticism [which are] ever busy and need feeding." Despite his fame, the folksy and caustic Darrow was not the ACLU's first choice to defend Scopes.

• The prosecution often invoked Southern pride in its arguments. As prosecutor Ben McKenzie proclaimed, "We don't need anybody from New York to come down here and tell us what [the Butler Act] means . . . The most ignorant man in Tennessee is a highly educated, polished gentleman [when compared] to the most ignorant man in some of our northern states."

• In each of his three campaigns for president, Bryan swept the South. However, he never carried Tennessee's Rhea County, the location of Dayton.

• It was not just Dayton that tried to cash in on the Scopes Trial. Although Bryan passionately defended religion at the trial, he also hired people in Dayton to promote Florida real estate for him during the trial.

• During his trial, Scopes received two washtubs of mail per day offering praise, condemnation, and the occasional marriage proposal. He and his friends piled it all in his yard and burned it.

• In reality, Scopes was available for the trial because he wanted to date "a blonde whose beauty and charm had caused [him] to linger in Dayton long enough to agree to the test case." Years later, Scopes admitted that the blonde deserved "as much credit as anyone for the trial's being held in Dayton that summer."

• *Baltimore Sun* reporter H.L. Mencken's coverage of the Scopes Trial helped transform the trial into a national event and is ranked as one of the top works of journalism of the 20th century.

• When publicity-seeking judge John Raulston started each day's proceedings with a Christian prayer, prosecutors bowed their heads while most of the defense team stared out the courtroom windows.

• On Sunday, July 26, the headline of Dayton's *Herald-News* asked, "Will William Jennings Bryan Ever Come Back?" (The article accompanying the headline had been written by Doris Stevens, the wife of Scopes's defender Dudley Field Malone.) That morning, Bryan made an unannounced appearance at Dayton's First Southern Methodist Church, and that afternoon, he died in his sleep in Dayton.

• Scopes's defense focused not on whether Scopes broke the law (i.e., taught human evolution), but instead on Bryan's perceived threat to individual liberty. Darrow *wanted* a guilty verdict; instead of trying to defend Scopes, he asked the jury to find his client guilty.

• John Scopes did not attend any part of his mismanaged appeal, noting that he "was not interested in the outcome and want to forget the entire episode." Unlike his trial, Scopes's appeal was described as "a flop of a story."

• No one—least of all John Scopes—suspected that the trial would become one of the most famous court cases in American history. But before it was over, the trial would test constitutional guarantees of liberties, such as speech and religion, and would pit science versus religion, academic freedom of teachers versus the academic freedom of students, and governmental authority versus teachers' rights.

Many issues over which Darrow and Bryan clashed in Dayton remain the topics of debate today. We still grapple with questions about who controls public education, as well as the extent to which citizens should decide what is taught in their public schools. What is the role of religion in public schools? What should teachers do with subjects that some may find offensive? Given the importance of these and related issues, it is not surprising that the Scopes Trial has been the topic of countless analyses, documentaries, and sermons and was later cited in numerous court decisions, including those issued by the US Supreme Court.

In this book, you will learn about the Scopes Trial, including its attendant issues, major players, causes, and consequences. You will encounter many entertaining characters, remarkable events, and multiple photographs and stories published here for the first time.

One

THE ROAD TO DAYTON

Evolution is the greatest menace to civilization in the world today.

—John Washington Butler,
the legislator who drafted Tennessee's antievolution law

As the crusade against evolution gained momentum, its leaders grew impatient for results. Riley's World Christian Fundamentals Association (WCFA), whose membership eventually exceeded five million, began several high-visibility campaigns to combat the teaching of evolution. In 1923, Oklahoma banned textbooks that included "the Darwin Theory of Creation versus the Bible Account of Creation." That same year in Florida, legislators declared it "improper and subversive to the best interests of the people" for public school teachers to teach evolution "or any other hypothesis that links man in blood relationship to any form of lower life." However, neither of these victories had a significant impact; the Oklahoma law applied only to grades in which evolution was not taught, and Florida's nonbinding resolution lacked the force of law.

Fundamentalists wanted a law banning the teaching of human evolution in public schools, and they got it in Tennessee. In early 1925, legislator John Washington Butler—who believed evolution was "the greatest menace to civilization in the world today"—introduced legislation banning the teaching of human evolution in Tennessee's public schools; violations were punishable by a fine of $100 to $500, a point that would become important during the appeal of Scopes's conviction. Butler's bill initially languished, and after a legislative recess in February, many observers believed the bill would die. But during that recess, fundamentalist preacher Billy Sunday's 18-day revival in Memphis drew more than 200,000 attendees, who heard Sunday denounce evolution as Satanic and evolutionists as "godforsaken cutthroats." When the legislature reconvened, it quickly passed Butler's bill and sent it to Gov. Austin Peay, who had earlier warned Tennessee that "something is shaking the fundamentals of the country." Harcourt Morgan, a biologist and the president of the University of Tennessee, feared that opposing Butler's legislation would hurt funding for the university, and he urged his faculty to remain silent about the bill. As he wrote to Governor Peay, "The subject of Evolution so intricately involves religious beliefs, concerning which the University has no disposition to dictate, that the University declines to engage in the controversy."

Although Peay wondered if Butler's legislation was necessary, he admitted that citizens "must have the right to regulate what is taught in their schools," and he signed the bill on March 21, 1925, in part to guarantee passage of another bill requiring teachers to read the Bible in class every day. Butler's bill, which became known as the Butler Act (chapter 27 of 1925's Public Acts of Tennessee), became the most famous of all the antievolution laws. Soon after its passage, Butler's legislation became the starting point of the Scopes Trial.

In 1859, British naturalist Charles Robert Darwin published *On the Origin of Species*, in which he proposes a mechanism for evolution, commonly known as natural selection. Although Darwin's book was controversial among the British clergy, it was relatively noncontroversial in the United States because the public was reassured by several leading scientists that Darwin's ideas were compatible with mainstream Christianity. (The American public was also preoccupied with events that would lead to the American Civil War less than two years later.) By the time he died in 1882, Darwin had written six editions of his famous book. Before long, some American theologians were blaming the teaching of evolution for a variety of societal ills. Forty-three years after Darwin's death, the Scopes Trial began in Dayton, Tennessee. (LC.)

ON

THE ORIGIN OF SPECIES

BY MEANS OF NATURAL SELECTION,

OR THE

PRESERVATION OF FAVOURED RACES IN THE STRUGGLE
FOR LIFE.

By CHARLES DARWIN, M.A.,
FELLOW OF THE ROYAL, GEOLOGICAL, LINNEAN, ETC., SOCIETIES;
AUTHOR OF 'JOURNAL OF RESEARCHES DURING H. M. S. BEAGLE'S VOYAGE
ROUND THE WORLD.'

LONDON:
JOHN MURRAY, ALBEMARLE STREET.
1859.

The right of Translation is reserved.

Although *On the Origin of Species* includes only one sentence about human evolution ("Light will be thrown on the origin of Man and his history."), contemporary readers understood that Darwin was replacing the notion of a perfectly designed and benign world with one based on an unending, amoral struggle for existence. Darwin offered no purpose for life except the production of fertile offspring. In Darwin's world, natural selection replaced divine benevolence as an explanation for adaptation. (LC.)

Dwight Lyman Moody was the first prominent American theologian to promote biblical inerrancy. Moody, a progenitor of fundamentalism, condemned theater, the disregard for the Sabbath, Sunday newspapers, and what he considered the atheistic doctrine of evolution. Moody, who avoided controversy, strongly influenced the fundamentalists who began the antievolution movement in America, including Billy Sunday and William Bell Riley. (LC.)

In 1919, Minnesota preacher William Bell Riley founded the World Christian Fundamentals Association (WCFA), the first and most formidable of the many early organizations to unite fundamentalists of all denominations in the United States. The WCFA sought to eradicate the teaching of evolution "not by regulation, but by strangulation," a campaign that Riley described as "a war from which there is no discharge." (FBC.)

Riley's WCFA was housed in First Baptist Church of Minneapolis, Minnesota, and by the early 1920s, Riley's speeches throughout the country condemning evolution were attracting thousands of people. On May 13, 1925, Riley invited William Jennings Bryan to represent the WCFA at John Scopes's upcoming trial in Dayton, Tennessee. The next day, Clarence Darrow volunteered to defend Scopes, thereby creating worldwide interest in the trial and prompting Scopes to conclude, "some Class-A sluggers were willing to champion our cause." (Both, FBC.)

EVOLUTION

Shall we continue to tolerate its teaching in our state schools?

Hear Dr. Riley

Pastor of the First Baptist Church
Discuss the above subject

SUNDAY 11 A. M. AT THE STATE THEATRE

The Fundamentals

A Testimony

Volume I

Compliments of
Two Christian Laymen

The Fundamentals is a series of 12 books published from 1910 to 1915 by the Bible Institute of Los Angeles (now Biola University). The books were given to "every pastor, evangelist, missionary, theology professor, theology student" and others to proclaim "a new statement of the fundamentals of Christianity." All costs were paid by California businessmen Lyman and Milton Stewart (the founders of Union Oil), who are described in the books as "two intelligent, consecrated Christian laymen." Fundamentalists, who opposed the adaptation of Christian theology to modern thought and secular society, believed in the inerrancy of the Bible, the Virgin birth, Christ's substitutionary atonement for sin on the cross, the physical resurrection, and Christ's eminent return. Although *The Fundamentals* do not focus on evolution, their publication became a symbolic reference point for the fundamentalist movement that, in 1925, produced the Scopes Trial in Dayton, Tennessee. (William McComas.)

By the early 1920s, famed baseball-player-turned-fundamentalist-evangelist William "Billy" Sunday was using his energetic, theatrical, and sometimes violent services to link evolution with prostitution, eugenics, and crime. Sunday, who claimed that Darwin was burning in hell, proclaimed that "education today is chained to the Devil's Throne" and that evolution was endorsed by "godless bastards and godless losers." William Jennings Bryan asked Sunday to testify at Scopes's trial, but Sunday declined. (LC.)

In the Northeast, the antievolution movement was led by militant fundamentalist John Roach Straton, the pastor of New York City's Calvary Baptist Church. Straton warned his congregation, "The great battle of the age is now on between Christianity and evolution . . . it is better to wipe out all the schools than undermine the Bible by permitting the teaching of evolution." This photograph shows Straton preaching in the streets of Manhattan. (American Baptist Historical Society.)

In the West, the fundamentalists' antievolution crusade was led by the flamboyant evangelist Aimee Semple McPherson. McPherson, whose marketing of her theatrical services rivaled even nearby Hollywood, attracted huge crowds to her church (Angelus Temple) in Los Angeles. Her most popular service featured depictions of Darwin, Hitler, and other villains, after which McPherson would emerge and read the national anthem. McPherson denounced evolution as "Satanic intelligence" that was "poisoning minds of children" and responsible for "jazz, crime, student suicides . . . and the peculiar behavior of the younger generation." McPherson, the first woman to deliver a sermon on radio, promised Bryan that 10,000 members of her church would be praying for his success in Dayton. In this photograph, McPherson uses her fist and a Bible to fight the apelike specter of Darwinian evolution. (Foursquare Gospel Church.)

In Texas and the South, the antievolution crusade was led by controversial J. Franklyn "Frank" Norris (left) of First Baptist Church of Fort Worth, Texas. Norris, the highest-paid preacher in the South, blamed the teaching of evolution for societal ills and urged his followers to "hang the apes and monkeys who teach evolution." Norris, who created many of today's unflattering stereotypes of fundamentalists, owned the *Searchlight*, a newspaper that reported Norris's and others' "war" against evolution. William Jennings Bryan (right) invited Norris to Dayton to help with Scopes's prosecution; Norris promised to be there but did not show up. (Left, LC; below, Arlington Baptist College.)

THE SEARCHLIGHT

Entered as Second-Class Mail Matter at the Postoffice at Fort Worth, Texas, March 15, 1917, Under the Act of March 3, 1897
PUBLISHED BY THE SEARCHLIGHT PUBLISHING CO.

VOL. VII.   FORT WORTH, TEXAS, FRIDAY, MAY 23, 1924.   NO. 26.

WAR DECLARED AGAINST EVOLUTION AT SAN ANTONIO

RILEY ASKS BAPTISTS IN NORTH TO OUST FEDERAL COUNCIL

By W. B. RILEY.

The Milwaukee Convention is immediately at hand. It is approached with mingled feelings. For the true progress marked in the past year, every Northern Baptist believer ought to be and is grateful. In the problems to be faced, these same men and women ought to be and are interested. It is not the purpose of this writer to review either the victories won or to call attention to the battle lost. The present situation in the Northern Baptist Convention offers many opportunities for praise and it equally affords special occasions for criticism. We shall engage in neither. Not that the first would not be a pleasant exercise, nor yet that the second might not produce profit, but that the past is fixed and represents the practically unalterable. Yesterday can never be the subject of vital interest on the part of progressive men. Tomorrow beckons too loudly to them, and toil they rightfully turn.

There may be attendants at the Northern Baptist Convention who have no unselfish interest in the progress of our great and common cause. Such delegates can never contribute ought to the organization or bring any blessing to the denomination. We are inclined to think they are few in number, and we can afford to forget them as inconsequential.

This writer, at least, has nothing against the denomination. Its name lent is more than forty years ago, from conviction. From that

NORRIS SENDS GENERAL BARTON MESSAGE

General Thomas D. Barton, Austin, Texas.

I have just read your address in today's Dallas News. Texas is facing the most serious crisis since the days of the Alamo and the one issue above every other issue is law enforcement.

I am in San Antonio holding big tabernacle evangelistic campaign; and return tonight. You are eminently correct in all that you say, therefore regardless of other issues and personal friends who are in the governor's race, please accept my support. The time has come when all law abiding citizens, ministers, fathers and mothers should be united on a great and heroic law enforcement program.

(Signed)
J. FRANK NORRIS.

THOMAS D. BARTON SOUNDS TOCSIN OF WAR

In his address beginning his campaign for governor, General Thomas D. Barton said among other fine things the following:

THE OFFICE OF TEACHING

(Stenographically Reported by L. H. Evridge.)

DR. NORRIS:

I want to give this morning a special message on the office of teaching. My text this morning is found in the 13th chapter and 1st verse of the Acts of the Apostles, the 13th chapter and 1st verse of the Acts of the Apostles.

"Now there were in the church that was at Antioch certain prophets and teachers; as Barnabas, and Simeon that was called Niger, and Lucius of Cyrene, and Manaen, which had been brought up with Herod the tetrarch, and Saul."

"Now there were in the church that was at Antioch certain prophets and teachers."

The first business of the New Testament prophets was to forthtell rather than to foretell. Now there was at Antioch certain prophets and teachers. Without doubt the church at Antioch influenced the whole world more than the church at Jerusalem. Barnabas was there, Paul was there and the Holy Spirit was the administrator of that church. Mark that—they didn't get orders from Rome or from Jerusalem or from New York or any where else. The Holy Spirit was their sole administrator. He called men into the ministry and called missionaries and sent missionaries forth. And that was the main reason that they had such a marvelous success in that day. Now the New Testament church had but one organization and that was the teaching, soul-winning organization and I will challenge any man to find where you have a thousand different side issues, wheels and machines in the church of the New Testament time. They didn't have it. They were too busy doing their main busi-

FROM THE PASTOR.

Rev. M. C. Bidson of San Antonio read the following letter from the Pastor Sunday morning and night:
To the Membership of the First Baptist Church:

As you well know I am in the most terrible fight of my life. The victory is already overwhelming and tremendous. Hundreds are being saved. Our altars say large were such crowds, such interest and such results in San Antonio.

The opposition has been fierce. All Rome and Hell have conspired against the meeting. But "if God be for us who can be against us?"

If ever in your whole life you prayed for a man, do it now. There is a great army of the Lord's heroic elect in San Antonio and they are on the firing line in the front line trenches.

I am holding up most splendidly, though have never worked so hard in all my life. I am going to stay until the victory is complete and decisive. I am preaching every day and night or do, toil, budget and redemption.

In great love,
J. FRANK NORRIS.

P. O. Box 246,
Orona, Texas, 5-12-24.
The Searchlight Pub. Co.,
Fort Worth, Texas.
Dear Sirs:

NORRIS IN GREATEST REVIVAL IN HISTORY OF ALAMO CITY

(From San Antonio Daily Light, Tuesday, May 20.)

NORRIS DENOUNCES EVOLUTIONISTS;
DECLARES WHOLE BIBLE IS INSPIRED

Does Not Mince Words in Attack on "Clique" Which Prevented Adoption of Resolution at Baptist Convention.

A short time ago the Fort Worth pastor received an invitation to come to the famous Metropolitan Tabernacle, Spurgeon's Church, in August. Since he has been in San Antonio, he has received the invitation from Dr. F. B. Meyer, minister of Christ's Church, London, to preach in the famous Kingsway Hall, which has a capacity of over ten thousand. The invitation asks him to speak on "the Advent Testimony, or "the Near Return of Our Lord."

The big tabernacle campaign has been characterized by one shock or explosion following in quick succession after another. Monday night, before a capacity audience, the evangelist, in clear, ringing voice, brought to his hearers a most terrific indictment against evolution on the theme "The Bible versus Evolution."

Boldly, and without mincing words, he launched an attack on the forces, the inside clique which prevented the adoption of the resolution that was offered at the Southern Baptist Convention, at Atlanta, against evolution, Dr. C. P. Stealey, of Oklahoma,

At the 1896 Democratic Convention in Chicago, William Jennings Bryan won the Democratic nomination for president after delivering his stirring "Cross of Gold" speech, which demanded an alternative silver-based currency to help cope with deflation caused by the excessive reliance on gold-backed money. The above photograph shows Bryan during the 1896 campaign, which he narrowly lost to William McKinley. (Bryan would later lose two more campaigns for the presidency, in 1900 and 1908, each time getting 45 to 48 percent of the popular vote). By 1912, Bryan had been appointed secretary of state by Pres. Woodrow Wilson (below, left). At this point, Bryan was not outspoken about evolution but suspected that it could undermine morality. Although Bryan was a powerful politician for decades and had served as a US congressman (from Nebraska), secretary of state was the most influential political office he ever held. (Both, LC.)

In 1924, Bryan visited Nashville, Tennessee, where he delivered his famed "Is the Bible True?" speech. Tennessee fundamentalists distributed copies of Bryan's antievolution speech to every state legislator, including second-term legislator John Washington Butler (shown here). Butler was a Primitive Baptist who believed that the United States was founded on the Bible, that evolution "is the greatest menace to civilization in the world" because it "robs the Christian of his hope," and that his law would "protect our children from infidelity." Butler drafted what became House Bill No. 185—better known as the Butler Act. The subsequent passage of the Butler Act made Tennessee the first state to ban the teaching of human evolution, thereby setting the stage for the Scopes Trial. (UTK.)

In 1921, John Scopes was a student at the University of Kentucky when its president, Frank LeRond McVey (shown here), led the battle against the first attempt to pass legislation banning the teaching of human evolution. Thanks largely to McVey's opposition, the legislation was defeated. After graduating in 1924 from the University of Kentucky with a bachelor's degree in "arts-law," Scopes accepted his first teaching job in Dayton.

Austin Peay IV (left) was the governor of Tennessee, who, on March 21, 1925, signed into law House Bill 185 (below)—the Butler Act—banning the teaching of human evolution in Tennessee's public schools. Like Bryan, Peay believed that "people must have the right to regulate what is taught in their schools" because "an abandonment of the old-fashioned faith and belief in the Bible is our trouble." Peay also wanted to garner legislative support for the subsequently passed An Act to Establish and Maintain a Uniform System of Publication (House Bill 780), which required teachers to begin each day of school with a Bible reading. When Peay signed Butler's bill, Bryan thanked Peay for "saving children from the poisonous influence of an unproven hypothesis." (Left, LC; below, RCHGS.)

HOUSE BILL NO. 185

BUTLER

Cl. 27

An Act prohibiting the teaching of the Evolution Theory in all the Universities, Normals and all other public schools of Tennessee, which are supported in whole or in part by the public school funds of the State, and to provide penalties for the violations thereof.

SECTION 1. BE IT ENACTED BY THE GENERAL ASSEMBLY OF THE STATE OF TENNESSEE, That it shall be unlawful for any teacher in any of the Universities, normals and all other public schools of the State which are supported in whole or in part by the public school funds of the state, to teach any theory that denies the story of the Divine Creation of man as taught in the Bible, and to teach instead that man has descended from a lower order of animals.

SECTION 2. BE IT FURTHER ENACTED, That any teacher found guilty of the violation of this Act, shall be guilty of a misdemeanor and upon conviction, shall be fined not less than One Hundred ($100.00) Dollars nor more than Five Hundred ($500.00) Dollars for each offense.

SECTION 3. BE IT FURTHER ENACTED, That this Act take effect from and after its passage the public welfare requiring it. Passed March 13, 1925.

SPEAKER OF THE HOUSE OF REPRESENTATIVES

SPEAKER OF THE SENATE

APPROVED:

March 21st 1925

GOVERNOR

Two

THE PATH TO JOHN SCOPES

Had we sought to find the defendant to present the issue, we could not have improved upon Scopes.

—Arthur Garfield Hays, one of John Scopes's defenders

John Thomas Scopes was born on August 3, 1900, on a farm in Paducah, Kentucky. He was the youngest of five children, and the only son, of Thomas and Mary Scopes. By the time he was 16, Scopes was living in Salem, Illinois, where he graduated from Salem High School in 1919. His school's convocation speaker was populist politician William Jennings Bryan. Six years later, Scopes and Bryan would meet again during Scopes's famous trial.

In 1924, Scopes earned a law degree (with a minor in geology) from the University of Kentucky and accepted a job at Rhea Central High School in Dayton, Tennessee. Scopes coached sports and taught math and science for $150 per month. Scopes was not hostile to religion, nor was he overly religious; as he noted, "I don't know if I'm a Christian . . . but I believe in God." While in Dayton, Scopes attended a Presbyterian church, but later admitted that he did so only to meet girls. He enjoyed teaching but bristled at governmental interference in the classroom because "once you introduce the power of the state—telling you what you can and cannot do—you've become involved in propaganda."

In April 1925, William F. Ferguson—the high school's biology teacher—was sick, and John Scopes took over his class for two weeks. According to his students, Scopes taught evolution and assigned the evolution chapter of the class textbook. Although he knew relatively little about evolution, Scopes accepted it because he considered it "the only plausible explanation of man's long and tortuous journey to his resent physical and mental development." Three months later, some of these students' testimonies became part of the prosecution's case against Scopes.

When Scopes began teaching in Dayton, the area's economy was struggling. In May 1925, a group of Dayton businessmen meeting at a local drugstore wondered if they could stimulate the area's fortunes by staging a "show trial" to test Tennessee's newly passed law banning the teaching of human evolution. They first had to find a defendant. Ferguson refused to get involved, so the businessmen shifted their attention to Scopes. They summoned Scopes from a tennis match to the drugstore; after he agreed to help the group by standing trial, Scopes returned to his tennis match. Scopes expected to be convicted, but he believed "the fine mess I got myself into" was "important for everyone in the country."

Roger Nash Baldwin was the first executive director of the American Civil Liberties Union (ACLU), a group chartered "to defend and preserve the individual rights and liberties guaranteed to every person in this country by the Constitution and laws of the United States." When Baldwin's secretary showed him a short newspaper article describing Tennessee's passage of the Butler Act, Baldwin began searching for a defendant there to challenge the law and to show that "a teacher may tell the truth without being thrown in jail." The Scopes Trial, on which the ACLU spent $8,993.01, was the group's first major trial. (ACLU.)

On page 5 of the May 4, 1925, issue of Chattanooga's *Daily Times* newspaper, an article titled "Plan Assault on State Law on Evolution" (right) caught the eye of Dayton engineer George "Rapp" Rappleyea. The next day, Rappleyea took the newspaper to the drugstore owned by F.E. "Doc" Robinson—often called "The Hustling Druggist" for his community activism. There, Rappleyea, Robinson, and other local businessmen conceived the idea of a trial to boost Dayton's struggling economy. (Dean Wilson.)

PLAN ASSAULT ON STATE LAW ON EVOLUTION

Civil Liberties Union to File Test Case.

The diminutive George "Rapp" Rappleyea (left) was an engineer who moved to Dayton in 1922 to coordinate the shutdown and sale of the failing Cumberland Coal and Iron Company. This company, which in 1925 consisted of six mines and 400 employees, was all that remained of the once-thriving Dayton Coal and Iron Company. By most accounts, Rappleyea was the lead instigator at the meeting at Robinson's Drug Store on May 5 where the idea for the Scopes Trial originated. After that meeting, Rappleyea signed the complaint that led to John Scopes's arrest for violating the Butler Act. (BC.)

By the late 1800s, the Dayton Coal and Iron Company (DCIC)—whose smokestacks are visible near the center of this photograph—was a primary economic engine of the Dayton area. Indeed, within the five years after DCIC began operations in the early 1880s, Dayton's population had increased from 300 to 2,700, and DCIC became one of the largest industrial developments in the South. In 1913, however, DCIC failed, after which the company began liquidation. By 1925, when the Scopes Trial started, Dayton's population had shrunk to about 1,800 people, many of whom were employed at Cumberland Iron and Coal Company, managed by George Rappleyea; this company was the remains of DCIC. The photograph above was taken from Sentinel Heights, with a view looking southwest toward Dayton. For scale, the smokestacks were nearly 200 feet high. (Both, UTK.)

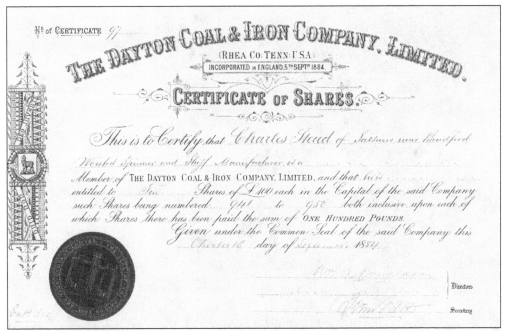

The declining fortunes of the coal and iron business played a critical role in the Scopes Trial. DCIC had been incorporated (above) in 1884 by Titus Salt and other English investors and by 1895 provided hundreds of jobs in the Dayton area. However, by 1910, the company was struggling and the area's economy was in trouble. This prompted local businessmen—led by mining executive George Rappleyea—to search for ways to improve the area's economy. In early May, Rappleyea and others met at the F.E. Robinson Company (commonly known as Robinson's Drug Store) and hatched their plan for stimulating Dayton's economy: a trial that soon became known as the Scopes Monkey Trial. Robinson's store (below), along with the 35-room Hotel Aqua next door, was a center of activity during the Scopes Trial. (Above, Randy Moore; below, SI Image No. SIA 2008-1102.)

Robinson had the exclusive concession to sell textbooks used at Rhea Central High School, which were displayed at the rear of his drugstore. One of those books, George Hunter's *A Civic Biology*, was used by John Scopes and became a focus of attention during Scopes's famous trial. (RCHGS.)

On May 5, 1925, Rappleyea and other Dayton businessmen met around this small table in Robinson's Drug Store to discuss how a trial to test the newly passed Butler Act could stimulate Dayton's economy. In this photograph, from left to right, mining engineer and trial instigator George Rappleyea poses with Walter White (superintendent of Rhea County Schools), druggist F.E. Robinson (standing), and Clay D. Green, White's assistant and a teacher who worked with Scopes. (BC.)

After the high school's regular biology teacher refused to get involved in the proposed trial, the trial's instigators shifted their focus to the school's science teacher and coach—John Scopes, who had been a substitute teacher in the biology class earlier that year. The instigators sent two students to find Scopes, who was playing tennis at the high school. After being asked by druggist and school board president F.E. Robinson if he would help out the town by standing for a test case, Scopes agreed and returned to his tennis match. When Scopes was later arrested, his case began; as Scopes noted, "I thought I'd lose my job, [but that] was nothing to lose." In this photograph, Scopes (right) discusses his predicament with Rhea County sheriff Robert "Bluch" Harris (left) in front of Robinson's Drug Store. (LC.)

Rhea Central High School hired John Scopes for $150 per month to teach math and science, and to coach football, basketball, and baseball during the 1924–1925 school year. This was Scopes's first and only year to work at the school. The popular Scopes, who considered coaching to be the most important part of his job, allegedly taught evolution when he substituted for the regular biology teacher in April 1925. The school was only three blocks from the Rhea County Courthouse, where Scopes would stand trial. (LC.)

John Scopes worked at Rhea Central High School, pictured above, which had opened in 1906 as Dayton's first public high school. The photograph below, taken in the summer of 1924, shows Scopes's football team. From left to right are (first row) Arch Shalton, Charles Stokley, Jack Hudson, C.L. Locke, and Dean Norman; (second row) Charles Hagler, Austin Tallent, Verdman Wells, Grady Purser, and Clair Elsea; (third row) John Scopes (coach), Henry Jones, Luther Welch, Edwin Williamson, and Carroll Tallent; (fourth row) Crawford Purser. Crawford Purser and Locke, who attended Scopes's trial every day (they usually sat in a windowsill of the courtroom, sometimes with WGN radio announcer Quin Ryan), were playing tennis with Scopes behind the high school when he was summoned to Robinson's Drug Store. (Both, BC.)

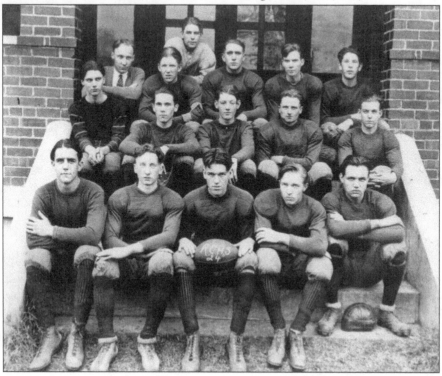

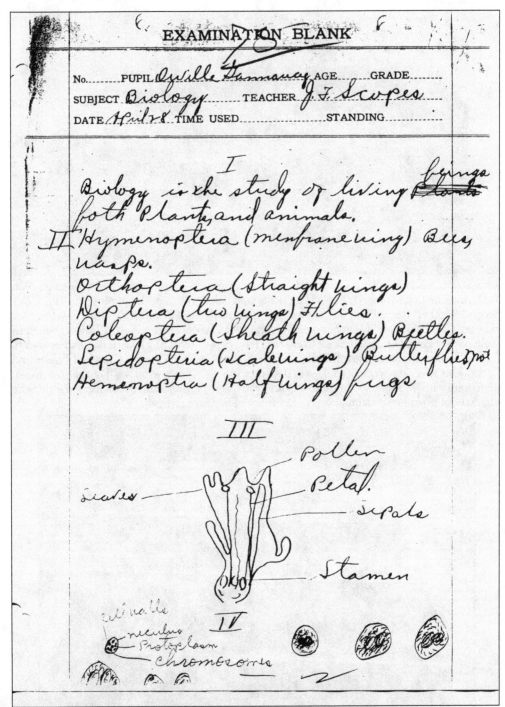

EXAMINATION BLANK

No......... PUPIL *Orville Hanmaucy* AGE......... GRADE.........
SUBJECT *Biology*......... TEACHER *J. F. Scopes*.........
DATE *April 28* TIME USED......................... STANDING.........

I

Biology is the study of living ~~plants~~ *beings*
both Plants, and animals.

II

Hymenoptera (menbrane wing) Bees
wasps.

Orthoptera (Straight wings)

Diptera (two wings) Flies.

Coleoptera (Sheath wings) Beetles.

Lepidopteria (scale wings) Butterflies moths

Hemenoptra (Half wings) bugs

III

Pollen

Petal.

Sepals

Stamen

Leaves

IV

nuclus

Protoplasm

chromosomes

When W.F. Ferguson, the regular biology teacher and school principal, got sick in late April 1925, John Scopes took over his class for two weeks. At the end of that time, Scopes gave an exam featuring questions about biology and major groups of living organisms (e.g., question 2 was about insects, and question 3 was about flowering plants). Scopes's exam included no questions that were specifically about evolution. (UTK.)

Three

THE CAST OF
CHARACTERS GROWS

I shall be pleased to be associated with your firm in the case . . .
I shall, of course, serve without compensation.

—William Jennings Bryan to Scopes's prosecutor Sue Hicks, May 20, 1925

Beyond Dayton, relatively few people took note of John Scopes's arrest and upcoming trial, until populist politician William Jennings Bryan—at the urging of William Bell Riley—agreed to help prosecute Scopes. Bryan's participation created immediate publicity and credibility for the trial.

More fame and notoriety for the trial came the next day when Clarence Darrow—the country's most famous defense attorney—announced that he would help defend Scopes. Darrow, who was famous for defending unpopular causes, had met Bryan at the Democratic National Convention in 1896, where he described Bryan's "Cross of Gold" speech as "the greatest oration that I ever witnessed." Darrow supported Bryan's presidential campaigns in 1896 and 1900, but by 1923, Darrow rejected Bryan's shift to religious fundamentalism. On July 4, 1923, a letter from Darrow posing 50 questions for Bryan appeared on the front page of the *Chicago Tribune*, but Bryan did not answer any of Darrow's questions. Realizing that John Scopes's trial could help him blunt Bryan's religious crusade, Darrow immediately volunteered to help defend Scopes in Dayton. Although several people in the ACLU objected to Darrow's involvement in the Scopes Trial, John Scopes settled the issue: "I want Darrow."

Bryan got to Dayton on Tuesday, July 7. He was one of the first out-of-town participants to arrive in Dayton, which was by then described as being "literally drunk on religious excitement." Soon thereafter, Bryan confronted Scopes, telling him, "You have no idea what a black and brutal thing evolution is." Bryan then began touting the upcoming trial as a religious and political war: "Teachers in public schools must teach what the taxpayers desire taught. Ramming poison down the throats of our children is nothing compared to damning their souls with the teaching of evolution. If evolution wins, Christianity loses."

When Darrow got to Dayton, he began responding to Bryan's claims: "The state of Tennessee doesn't rule the world yet. Scopes is here because ignorance and bigotry are rampant, and [that] is a mighty combination. Had we Mr. Bryan's ideas of what a man may do toward free thinking, we would still be hanging and burning witches, and punishing persons who thought earth was round."

Most of the verbal jousts between Bryan and Darrow were made indirectly through the press, but people attending the trial were looking forward to a face-to-face confrontation.

EVOLUTION CASE STARTED AT RHEA

Dayton Professor · to Make Test of New Law.

New York Concern to Supply Finances—Prof. Scopes Arrested by Friends.

On May 6, the *Chattanooga Daily Times* announced that John Scopes had been arrested "by friends" (but never detained) and that Rhea County would host a trial challenging the Butler Act. (The "friend" who actually arrested Scopes on May 5 was deputy Perry Swafford.) The newspaper later denounced the trial as "a humiliating proceeding" that embarrassed every lawyer in Tennessee. (UTK.)

The Scopes Trial generated thousands of cartoons and other editorial statements across the world. This cartoon—complete with a monkey collecting money—shows how Dayton tried to use the Scopes Trial to generate publicity and stimulate its struggling economy. (UTK.)

When Bryan arrived in Dayton for the Scopes Trial, he was 65 years old and his health was failing. His had diabetes, heart trouble, and an enormous appetite that had made him overweight. After Scopes's trial, Bryan visited a physician in Chattanooga for a checkup. As had happened in the past, Bryan was told to rest, but he seldom did so. (LC.)

At Royal Palm Park just outside of Miami, Florida, Bryan (left) taught a Sunday school class with 5,000 members. There, and elsewhere, Bryan blamed many societal ills on the teaching of evolution. In 1923, the Florida Legislature passed a resolution written by Bryan claiming the teaching of evolution was "improper and subversive to the best interest of the people." (BC.)

HEAR
HONORABLE WILLIAM
JENNINGS BRYAN
At the State Fair
HIPPODROME
(TAKE COMO AVE. STREET CAR)

Sunday Oct. 22nd At 2:30 P. M.

Subject
"EVOLUTION"
A Menace to Christianity Education and Civilization

MASSED CHOIR
650 VOICES
Under Direction of F. V. Steel Will Sing

Doors Open At 2 P. M.   **ADMISSION FREE**

8000 Seats. 3000 Reserved for Students

AUSPICES OF NORTHWESTERN BIBLE AND MISSIONARY TRAINING SCHOOL, MINNEAPOLIS

By 1924, Bryan's speeches condemned evolution as a "poison" responsible for World War I and "all the ills from which America suffers." Bryan's speeches attracted huge crowds; for example, the speech advertised here—"Evolution: A Menace to Christianity Education and Civilization"—attracted more than 10,000 people in Minneapolis and branded evolution the intellectual source of moral disorder. Bryan believed the Scopes Trial would be a turning point in Christianity. (FBC.)

Chicago lawyer Clarence Darrow had supported Bryan's early presidential campaigns but by 1923 was disappointed by Bryan's shift to religious fundamentalism. When Bryan volunteered to help prosecute John Scopes, Darrow volunteered to defend Scopes the following day. Although Darrow had served without fee in many cases, the Scopes Trial was the only trial in which he volunteered his services. After spending $2,000 of his own money on the trial, Darrow claimed, "I never got more for my money." (BC.)

The Scopes Trial attracted preachers, hustlers, and others to Dayton, including "John the Baptist III," "Deck Carter, The Bible Champion of the World," and Lewis Levi Johnson Marshall, who proclaimed himself to be "The Absolute Ruler of the Entire World, without Military, Naval or Other Physical Force." John Scopes described the gathered masses as "con men and devotees" who, combined with a variety of "pulpit heroes" and "sideshows and freaks," produced "one of the rarest collections of screwballs I have ever seen in my life." This photograph shows an itinerant evangelist preaching to a crowd near the courthouse. (BC.)

Banker A.P. "Mr. Dayton" Haggard was the chief commissioner of Dayton—and one of Dayton's most prominent civic leaders—during the Scopes Trial. Haggard, the father of prosecuting attorney Wallace Haggard, was born Pleasant Andrew Haggard but later reversed his first two names. Haggard, who helped prepare Dayton for the Scopes Trial, had one of Dayton's most illustrious careers, but he ended up broke (a victim of the Great Depression) and committed suicide in F.E. Robinson's home. (RCHGS.)

Civic leaders in Dayton wanted to impress the throngs of visitors anticipated for Scopes's trial. To help accommodate the crowds, A.P. Haggard organized a Committee on Reservations, consisting of, from left to right, Clarence E. Toliver, S. Carl Patton, B.M. "Burt" Wilbur, and chairman W.N. "Bill" Morgan. (UTK.)

Dayton businessmen F.E. Robinson and W.E. Morgan prepared a 28-page booklet titled *Why Dayton, of All Places?* to help lure investors and others to Dayton. The booklet, which includes quotations from Sinclair Lewis's *Main Street* while alluding to various religious, political, and philosophical issues associated with the Scopes Trial, is—according to the booklet itself—a "half playful, half serious" attempt to "start something and maybe it would be interesting." This photograph shows the hatless Robinson unloading packages of the booklet when they arrived in Dayton. (Both, RCHGS.)

In the hills outside of Dayton (above), enthusiastic members of the Holiness Movement held evening camp services during the Scopes Trial that included baptisms and ecstatic celebrations. These celebrations often included shouting, frenetic movements, and rolling on the ground, which led to their pejorative name "Holy Rollers." The group shown here was led by preacher Joe Leffew. Meanwhile, in town (below), Mississippi evangelist Thomas Theodore "T.T." Martin's book *Hell and the High School* sold briskly on Dayton's streets and in Robinson's Drug Store. In that book, Martin, an officer in William Bell Riley's Anti-Evolution League of America, claims that Germans who poisoned children were "angels" compared to teachers who teach "the deadly, soul-destroying poison of Evolution . . . [The teaching of evolution] is the greatest curse that ever fell upon this earth." (SI Image No. 2008-1132, RCHGS.)

During the Scopes Trial, visitors walking down Market Street were undoubtedly surprised to encounter Darwin's General Mercantile (above), which advertised clothes that were the "fittest" and that "Darwin Is Right—Inside." The "Everything to Wear" store was operated by James Robert "Red Jim" Darwin (1866–1939, far left, above and below) and his wife, Margaret (Sharp) Darwin (second from left, above), who boarded two reporters in their home (today's Broyles-Darwin House at 108 East Idaho Avenue). In an interesting twist, Red Jim was a descendant of William Darwin, the great-grandfather of British naturalist Charles Darwin. (Both, UTK.)

Robinson's Drug Store, where the Scopes Trial originated, was a center of activity before, during, and after the trial. During the Scopes Trial, F.E. Robinson—for the first time ever—opened his drugstore on Sundays to serve the crowds of people in town for the trial. Robinson, who promoted himself as "The Hustling Druggist," continued to tout his (and his drugstore's) association with the Scopes Trial until his death in 1957. (BC.)

Reporters from around the world converged on Dayton to cover the trial; they wired more than 160,000 words each day about the proceedings on more than 10 miles of new cable laid specially for the trial. Although most of the reporters were men, the woman shown here—Nellie Kenyon—was a reporter for the *Chattanooga News* who later rose to fame with her interviews of luminaries such as Scopes, teamsters boss Jimmy Hoffa, and Al Capone. (RCHGS.)

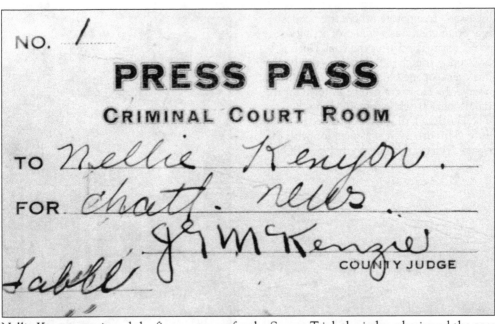

NO. 1

PRESS PASS

CRIMINAL COURT ROOM

TO _Nellie Kenyon._

FOR _chatt. news_.

J. G. McKenzie

COUNTY JUDGE

Isabell

Nellie Kenyon was issued the first press pass for the Scopes Trial; the judge who issued the pass was Rhea County judge James Gordon McKenzie, the son of Scopes's prosecutor Ben McKenzie. Kenyon also wrote the first story about Scopes's arrest. The initial press coverage, which was fed to more than 2,000 newspapers worldwide, had an ongoing influence on the public's opinion of the Scopes Trial, Dayton itself, and even the South. (Dean Wilson.)

The army of reporters in Dayton included Henry Louis "H.L." Mencken, a satirical iconoclast who worked for the _Baltimore Evening Sun._ Mencken, who urged Darrow to volunteer to defend Scopes, coined the terms "Bible Belt" and "Monkey Trial" and enraged Dayton's residents when he referred to them as "yokels," "morons," and "gaping primates." Like most reporters, Mencken left Dayton before Darrow's famous questioning of Bryan on the courthouse lawn. (LC.)

This is a view looking east along Dayton's Main Street during the time of the Scopes Trial. Robinson's Drug Store, where the idea for a trial originated, is on the right, just beyond the three-story Hotel Aqua. (BC.)

Most visitors and trial participants arrived in Dayton aboard trains, some of which were specially chartered for the event and adorned with advertisements urging "Let's All Go to the Scopes Trial!" (SI Image No. SIA 2008-1124.)

Bryan arrived in Dayton on July 7 and was greeted at the train station by a crowd of 250 people. Soon after departing the train, Bryan was given telegrams from evangelists Aimee Semple McPherson and Billy Sunday, both praising Bryan for his defense of fundamentalists' ideals. Later that week, Bryan told John Scopes, "You have no idea what a black and brutal thing this evolution is." Bryan's arrival in Dayton calmed the town's circus-like atmosphere, but intensified the war of words between Scopes's prosecutors and defenders. (UTK.)

The crowd awaiting Bryan at the train station included fellow prosecutors Ben McKenzie (to the right of Bryan, in dark suit with arms folded) and Herbert Hicks (in white suit, holding his hat and Bryan's bag). Soon after arriving, Bryan proclaimed that he was in Dayton to defend religion. (BC.)

After arriving in Dayton, Bryan and fellow prosecutors went to the home of Frederick Richard Rogers on Market Street, where Bryan stayed during and after the trial. This photograph shows the group in front of the Rogers' home, which was described as being "in the best residential section" of Dayton. From left to right are Wallace Haggard, Gordon McKenzie, F.R. Rogers, Bryan, Sue Hicks, Herbert Hicks, and Ben McKenzie. (BC.)

During the Scopes Trial, Bryan, his wife, Mary, and their entourage rented (for $25 per week) the home of Frederick Richard Rogers, a pharmacist at Robinson's Drug Store. This photograph shows Bryan holding young Jean Rogers in the Rogers' yard on July 7, just three days before the trial started. Rogers, a founder and trustee of Bryan College, died at a dinner at the college in 1964. (Pat Hawkins Guffey.)

As the beginning of the Scopes Trial approached, Bryan gave sermons and speeches throughout the Dayton area. In those speeches, Bryan described the upcoming trial "not a joke but an issue of the first magnitude" that would expose the "gigantic conspiracy among atheists and agnostics against the Christian religion." In this photograph, Bryan pauses to cool himself with a fan advertising the F.E. Robinson Company. (BC.)

Mary Baird Bryan, the wife of William Jennings Bryan, arrived in Dayton on July 9 and was in court every day of the Scopes Trial. Mary privately objected to her husband's crusade against evolution but was by his side in Dayton. Mary suffered from arthritis, which confined her to a wheelchair during the trial. After the Scopes Trial, Mary completed her husband's memoirs. When several towns vied to be the site of William Jennings Bryan University, Mary settled the issue by urging planners to build the university in Dayton. (LC.)

```
                              July ₤ 6, 1925
Hicks Bros.
Dayton, Tenn.


The Progressive Dayton Club will entertain
Hon. William Jennings Bryan with a Banquet
on Tuesday evenihg at 8:00 o'clock,
                Hotel Aqua.


Present this card for admittance.
```

Wallace O. Haggard
                                Secretary

Dayton's Progressive Club—a group of local leaders that included several of the trial's instigators—budgeted $5,000 to promote Dayton and accommodate visitors during Scopes's trial. The club also hosted banquets for Bryan and Darrow when each arrived in town. At the banquet for Bryan at Hotel Aqua on July 7—an event that John Scopes described as "the social event of all of Dayton's history"—Bryan sat with Scopes and John Randolph Neal and repeated his proposal for a constitutional amendment banning the teaching of human evolution. Invitations to the banquet were signed by Wallace Haggard, who helped prosecute Scopes. (Above, UTK; below, RCHGS.)

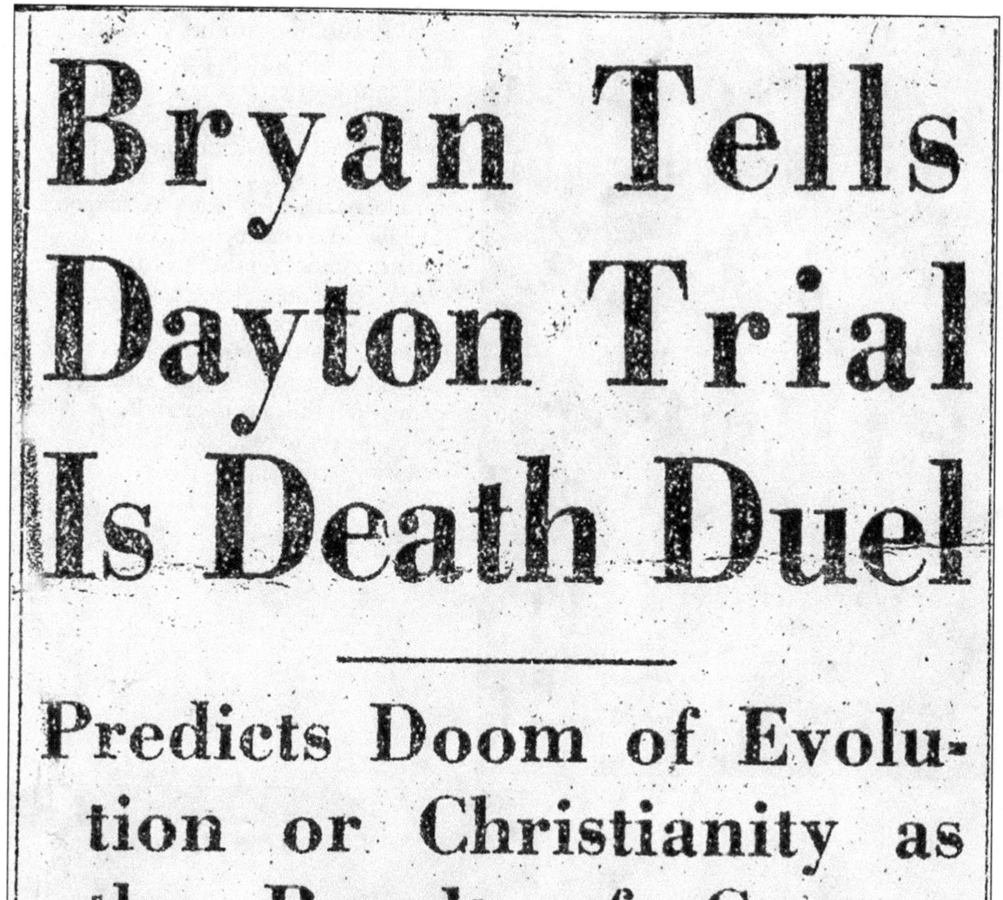

Bryan Tells Dayton Trial Is Death Duel

Predicts Doom of Evolution or Christianity as the Result of Contest and Is Eager for Fray

---

Scores Triumph as He Arrives in Town

During his speech to the Dayton Progressive Club, Bryan declared Scopes's upcoming trial a "duel to the death" between the Bible and infidelity, adding "if evolution wins in Dayton, Christianity goes . . . for the two cannot stand together. They are as antagonistic as light and darkness, as good and evil." This headline is from the July 8, 1925, issue of the *New York Herald Tribune*. (BC.)

Tom Stewart, the attorney general for Tennessee's 18th Judicial District, led Tennessee's prosecution of John Scopes. Stewart argued that Scopes's case involved no religious questions, but on the fourth day of the trial, he introduced the King James Version of the Bible into evidence as Exhibit 2. Stewart led most of the legal arguments for the prosecution. Except for Darrow's questioning of Bryan on July 20 (which Stewart opposed), Stewart controlled the Scopes Trial. Stewart's strategy left the defense with little choice except to request a conviction that they would appeal. (RCHGS.)

This photograph shows prosecutor Ben McKenzie wiping the sweat from his forehead outside the courthouse on the fourth day of proceedings. At the time of the Scopes Trial, McKenzie—who had been the first attorney general of the newly created 18th Judicial District—was one of the most prominent attorneys in Tennessee. After the trial, McKenzie—who believed that anyone who accepted evolution was an atheist or agnostic—gave harsh antievolution speeches at Chattanooga's Centenary Methodist Church and elsewhere. McKenzie and Darrow became close friends. When McKenzie was arrested in February 1926 for violating liquor laws, Darrow offered to defend him. McKenzie declined the offer. (RCHGS.)

Dayton attorney Sue Hicks raised money for the prosecution and helped prosecute John Scopes. Years after the trial, Hicks claimed that the idea for the trial originated in his office during a discussion with school superintendent Walter White. Hicks later defeated White in a race for the Tennessee legislature, after which he served for 22 years as a judge. Hicks became a folk hero when he inspired Chicago writer Shel Silverstein to write "A Boy Named Sue," a song popularized in 1969 by Johnny Cash's live recording *At San Quentin*. (BC.)

Defense attorney Clarence Darrow was also warmly greeted by a crowd when he arrived in Dayton. In this image, taken at Dayton's train station, Darrow (in hat) shakes hands with John Scopes as co–defense attorney John Neal (between Darrow and Scopes) looks on. Soon after this photograph was taken, Darrow described William Jennings Bryan as "the idol of all Morondom." (RCHGS.)

Dudley Field Malone was the only professing Christian on Scopes's defense team. On the trial's fifth day, Malone pointed out that the world's interest in the Scopes Trial was not about "whether John Scopes taught a couple of paragraphs" from the course textbook, but instead was about the religious agenda that Bryan was in Dayton to defend; this made it virtually impossible for either side to return to the question of what Scopes had actually taught. On that same day, Malone delivered his "Duel with the Truth" speech, which Scopes later described as "the most dramatic event I have attended in my life" and which Bryan declared "was the greatest speech I have ever heard." Years later, Malone admitted that his famous speech in Dayton was the only extemporaneous speech he ever gave. (LC.)

Scopes's defense was headed by John Randolph Neal, who ran a proprietary law school in Knoxville. One newspaper described the disheveled Neal as looking "as if he had just come from beneath a box car." On July 3, Neal tried unsuccessfully to transfer Scopes's trial to federal court so the defense could argue the constitutionality of the law. Many people at the ACLU opposed Clarence Darrow being on the defense team. As chief counsel, Neal could have removed Darrow from the team, but he did not do so. (RCHGS.)

The Scopes Trial took place in the second-floor courtroom of Dayton's Rhea County Courthouse, a Romanesque/Italian villa–style courthouse built in 1891. The 65-foot-by-65-foot courtroom, which is still in use, remains the largest courtroom in Tennessee. During Scopes's trial, Chicago's WGN radio station set up speakers outside the courthouse so that overflow crowds could hear the courtroom's events. In 1997, the courthouse was designated a National Historic Landmark. In its basement is the Rhea Heritage and Scopes Trial Museum, which displays exhibits and memorabilia related to the trial and Rhea County. (RCHGS.)

At the time of Scopes's trial, the Rhea County Courthouse was surrounded by a cast-iron fence that visitors negotiated with steps that were interspersed around the courthouse square's perimeter. This photograph shows Maude Thomison and Max Ramsey sitting atop one of the "bridges" over the fence. Maude was the fashion-conscious daughter of Dr. Walter Fairfield Thomison, the attending physician at Bryan's death in Dayton a few days after the trial. During the Scopes Trial, Maude dated Boston reporter W.A. Macdonald. (Pat Hawkins Guffey.)

Before the start of the Scopes Trial, Dayton officials had planted these young trees beside the Rhea County Courthouse. Beyond these trees, and just outside the courthouse, is the platform at which Darrow would famously question Bryan in the climactic event of the Scopes Trial. The trees shown in this photograph are now part of the Scopes Trial Grove on the courthouse square. (UTK.)

Judge John Tate Raulston of Tennessee's 18th Judicial District, his wife, Eva (right), and their two daughters came to Dayton from Winchester, Tennessee, and stayed at the Hotel Aqua during Scopes's trial. Raulston was an ordained minister who believed that God wanted him to preside at Scopes's famous trial. Raulston, who had been elected judge in 1918 and whose trials often involved bootlegging, later claimed that Dayton "was like a beehive when I arrived." (UTK.)

When the Scopes Trial finally started, four-year-old Thomas Jefferson "Tommy" Brewer drew the names of jurors out of a hat while seated on Judge Raulston's bench. Before he died in 2003, Tommy was the last surviving participant in the Scopes Trial. (BC.)

The jury in the Scopes Trial, shown here with the sheriff and judge, was excused for most of the trial and heard little of the trial's discussions. Pictured are, from left to right, (first row) W.G. Taylor, Jesse B. Goodrich, Jack R. Thompson (foreman, with white moustache and goatee), William "Bill" Day, Robert Lee Gentry, John S. Wright, and Judge Raulston (far right); (second row) Rhea County sheriff Robert B. "Bluch" Harris, R.L. West, William Davis "Billy" Smith, James W. Riley, John W. Dagley, and John Hayes Bowman. Absent from the image is juror W.F. Roberson. (BC.)

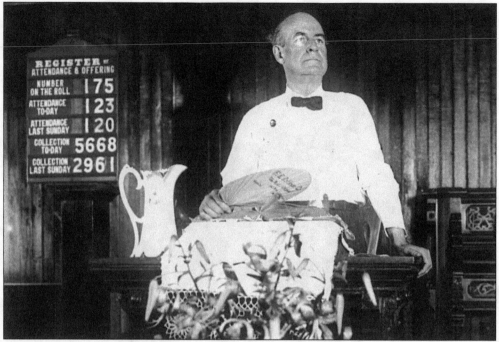

At services on the morning of Sunday, July 12, at Dayton's First Southern Methodist Church, Bryan was introduced by Rev. Charles R. "Parson Charlie" Jones as an "ambassador of Christ." Bryan then preached for 70 minutes from the pulpit and led a prayer. Judge Raulston and his wife and two children sat in the front row. (BC.)

After the service, Bryan posed with worshipers outside First Southern Methodist Church. This image shows just a part of the yearbook-like photograph of that gathering. Two weeks later, Bryan would make his last public appearance at this church. (First Methodist Church, Dayton.)

During the trial's first weekend, Bryan spoke to a large crowd from a platform erected on the side of the Rhea County Courthouse. On July 20, Bryan would be questioned by defense attorney Clarence Darrow on this same platform in the Scopes Trial's climactic event. (BC.)

July 15 was the trial's fourth day of proceedings, and jury members—all but one of who belonged to a local evangelical church—were sworn into duty. Three members of the jury reported having read no book except the Bible, and another juror could not read. (BC.)

The afternoon issue of the July 14, 1925, *New York Journal* announces the courtroom exchanges resulting from Clarence Darrow's objections to Judge Raulston's decision to open each day's proceedings with prayers. Darrow, who objected to Raulston "turning court into a meeting house," was overruled.

Four

THE TRIAL OF THE CENTURY BEGINS

*Nothing will satisfy us but for broad victory, a knockout which
will . . . prove that America is founded on liberty and not on narrow, mean,
intolerable and brainless prejudice of soulless religio-maniacs.*

—Clarence Darrow, John Scopes's defender

As the Scopes Trial approached, civic leaders of Dayton began preparing for what they hoped would be a flood of visitors to their struggling town. Numerous residents rented their homes to visitors (including to trial principals Clarence Darrow and William Jennings Bryan), local hotels increased their fees, and town officials hurriedly built temporary bathrooms, cleaned sidewalks, and mowed lawns to spruce up Dayton. Vendors began stockpiling monkey-related souvenirs and evolution-related books, and religious leaders added services for the faithful.

Chicago radio station WGN made history by sending Quin Ryan—an announcer famous for broadcasts that were advertised as being "almost as good as being there"—to Dayton for the first live national broadcast of a trial in American history. WGN, which spent $1,000 per day on the broadcasts, also used in-court microphones and outdoor speakers to broadcast proceedings to the trial's overflow crowds. The freshly painted courtroom was even rearranged to accommodate WGN's microphones atop the station's four-foot-high wooden stands. (These microphones and transmission wires appear in many of the trial photographs.) Ryan, who was assisted by Dayton youngster Carmack Waterhouse, often sat in a windowsill as his radio station broadcast the trial's proceedings. Ryan, like Scopes, Darrow, and Bryan, was a celebrity in Dayton during the trial.

Other vendors also prepared for the big event. American Telephone & Telegraph laid more than 10 miles of cable, Western Union added 22 on-site telegraphers, and Southern Railway added extra service to Dayton to meet customers' demands. More than 150 journalists arrived for the trial (from as far away as Hong Kong) and produced more than two million words of coverage; indeed, more words about the trial were cabled to Europe and Australia than had ever been cabled for any other event. To feed the public's insatiable appetite for news about Scopes's trial, several reporters fabricated some of their stories.

Dayton, which had become a media circus, was flooded with a motley assortment of vendors, circus performers, street salvationists, sincere spectators, money-grubbing hucksters, and—as noted by several newspapers—other "cranks and freaks." When more than 500 visitors showed up for the trial's opening day, Dayton was ready to cash in on its new fame. The Scopes Trial—which the July 20, 1925 issue of *Time* magazine would describe as "the fantastic cross between a circus and a holy war"—was ready to begin.

Although Scopes's trial included eight days of proceedings, the jury was in the courtroom for only about three hours during the entire trial. This photograph shows the jury on the courthouse lawn, where jurors spent much of their time while lawyers in the courtroom debated procedural issues associated with Scopes's trial. (BC.)

John Scopes's 65-year-old father, Thomas Scopes (right), came from Paducah, Kentucky, to Dayton to support his son. The elder Scopes, who was a railroad machinist and union activist, believed that the upcoming trial was his son's chance to serve his country. This photograph shows father and son sitting on the porch of the *Chattanooga News* Newspaper Club. (RCHGS.)

Rev. Howard Byrd (right) pastored two churches in Dayton—First Avenue Methodist Episcopal Church and Five Points Methodist Episcopal Church. When the 35-year-old Byrd, who led a prayer at Scopes's first hearing on May 25, invited pro-evolution preacher Charles Francis Potter—pastor of New York's West Side Unitarian Church—to speak about evolution at his First Avenue church on July 12, his congregation was enraged. Potter, a former fundamentalist, was a vocal opponent of Bryan and expected to testify for the defense, but instead became a freelance writer at the trial. Just before the Scopes Trial started, trial instigator George Rappleyea (left)—a member of Byrd's Five Points church—credited Byrd as being the "inspiration" for the trial. (UTK.)

The image above shows, from left to right, George Rappleyea, Methodist preacher Howard Byrd, and Unitarian preacher Charles Potter in front of Dayton's First Avenue Methodist Episcopal Church, where Byrd was pastor. Byrd wanted both sides of the evolution controversy to be heard by his congregation during the trial. When he invited Potter to preach at his church, his congregation threatened a boycott and demanded that Byrd rescind the invitation. Byrd refused, instead proclaiming, "I have quit. I have not resigned; I have quit." The image below shows, from left to right, Unitarian minister Leon Birkhead (who came to Dayton from Kansas City to help the defense), Howard Byrd, George Rappleyea, an unidentified man holding Byrd's son John, Frank Thone (of the Science Service), and Charles Potter holding Byrd's daughter Lilian. (Above, SI Image No. 2008-1096; below, SI Image No. 2008-1095.)

Edwin Emery Slosson (right) was a self-described science "renegade" who in 1921 became the first director of the newly founded Science Service, a nonprofit organization devoted to popularizing science by marketing syndicated articles to magazines and newspapers. Slosson chose 29-year-old managing editor Watson Davis to lead Science Service's coverage of the Scopes Trial. (SI Image No. 2007-0014.)

In Dayton, Watson Davis (left) worked as a journalist, but he also helped the defense team choose its expert witnesses. On July 17, Davis declared himself part of the "scientific group"; he also arranged for all of the scientists to sign a letter endorsing Darrow's "ability, high purpose, integrity, moral sensitiveness, and idealism." At the trial, Davis sat by H.L. Mencken, immediately behind John Scopes and the defense team. Davis later helped start the Westinghouse Science Talent Search and was a founding member of the National Association of Science Writers. (SI Image No. 2016-0428.)

Darrow and his wife, Ruby, stayed at the home of Luther Morgan, which they rented for $500 for "however long the trial lasts." Luther's son, 14-year-old Howard "Scrappy" Morgan, was the first of Scopes's students to testify at Scopes's trial. When asked by Clarence Darrow if learning about evolution had "hurt you any" or caused him to leave the church, the young Morgan responded, "No, sir." (Richard Cornelius.)

While street preachers met the crowd's spiritual needs, temporary bathrooms such as these—both "For Ladies"—behind the courthouse helped Dayton residents meet the sanitary challenges posed by the crowds. Wild-eyed antievolution crusader T.T. Martin adorned these and other sites in Dayton with signs urging people to "Read Your Bible." (BC.)

In this photograph, John Scopes (left) walks by one of T.T. Martin's "Read Your Bible" signs with his attorney John Neal (center) and trial instigator George Rappleyea. Other signs posted throughout Dayton proclaimed: "You Need God in Your Business," "God is Love," and "Where Will You Spend Eternity?" During Scopes's trial, the defense team objected to the "Read Your Bible" signs, as they did to the prayers that opened each session of court. (BC.)

Emanuel "E." Haldeman-Julius was a socialist, atheist, and publisher who arrived in Dayton on July 10 after driving hundreds of miles with his wife, Anna Marcet, to witness the Scopes Trial. Haldeman-Julius published hundreds of titles in the Little Blue Book series discussing topics such as science, sex, politics, and religion (which he describes as "bunk"). The books sold briskly in Dayton during the trial. During proceedings, Bryan cited one of Haldeman-Julius's books that describes Darrow's defense of thrill killers Leopold and Loeb. The photograph at left shows Haldeman-Julius in front of the "Mansion" where defense witnesses stayed. (SI Image No. 2008-1092.)

LITTLE BLUE BOOK NO. 1424
Edited by E. Haldeman-Julius

The Famous Examination of Bryan at the Scopes Evolution Trial

Clarence Darrow

This Little Blue Book published by Haldeman-Julius documents Darrow's questioning of Bryan at the Scopes Trial. At the time of his death in 1951, Haldeman-Julius had published more titles and volumes than any other company in the world. (William McComas.)

JOE MENDI IN DAYTON TENN. ON "MONKEY BUSINESS"

Poor mendi died the second week he was in Dayton the Climate was too warm or ex cite ment too keen

Outside the courthouse, one of the most popular attractions during the Scopes Trial was Joe Mendi, a chimpanzee brought to Dayton for "monkey business" by circus owner Lew Backenstoe. People paid 25¢ to shake Joe's hand. Mendi was billed as "The $100,000 Chimpanzee with the Intelligence of a Five-Year Old." Although this entry in a Dayton scrapbook claims that Joe died during the second week of Scopes' trial, Joe—who was insured for $100,000—did not die until 1930. Before and after the Scopes Trial, attendance at zoos through Tennessee increased as visitors flocked to see monkeys and apes. (BC.)

Joe made several appearances throughout Dayton during the Scopes Trial. In this photograph, Joe gives a piano concert for Clark Robinson (F.E. Robinson's wife) and some students. After the Scopes Trial, Joe's handlers often reminded crowds that Joe was at Dayton, where "he was asked to appear, but he did not testify." (BC.)

During the Scopes Trial, Dayton's merchants sold a variety of books, pennants, and monkey-related products and souvenirs. The items most popular with girls and women were monkey dolls wearing liberty caps. This photograph was taken during the trial near the courthouse. (UTK.)

In this street scene during the Scopes Trial, Marguerite Purser (left) and Andrewena Robinson (daughter of F.E. Robinson) pose with their dolls. Both Purser and Robinson were friends of John Scopes's. (BC.)

In early April, Rhea County clerk E.B. Ewing ordered Rhea County sheriff Robert Harris to summon five of John Scopes's students—Howard Morgan, James Benson, Morris Stout, Traynor Hutcheson, and Jack Hudson—to appear in court on July 10, the opening day of the Scopes Trial. Of these students, only Morgan eventually testified at the trial. (UTK.)

The Scopes Trial was the first trial in US history to be broadcast live nationwide on the radio. In this scene from court proceedings (above), the microphone of Chicago's WGN radio station is prominent in the left foreground. Standing to the right of the microphone are prosecutors Wallace Haggard (with face blurred), Herbert Hicks, and defense attorney Dudley Field Malone (in coat with arms folded). WGN's coverage in Dayton was headed by Quin Ryan. During Scopes's trial, Ryan—a celebrity in Dayton because of the new technology that accompanied him—stayed in F.E. Robinson's home. One of the four original WGN microphones (left) can be seen today at Bryan College. (Above, BC; left, William McComas.)

On the fourth day of proceedings, Clarence Darrow (left, coatless) addressed the jury, which is seated directly in front of him. In the left foreground, with his back to the camera, Dudley Malone sits by a WGN microphone. (BC.)

Many parts of the Scopes Trial's proceedings were tedious. In this photograph, William Jennings Bryan (just right of center) yawns as he fans himself to stay cool. To the left of Bryan are Sue Hicks, Herbert Hicks (partially obscured), Ben McKenzie (with his left hand to his head), Wallace Haggard, and Gordon McKenzie. In the lower right corner is lead prosecutor Thomas Stewart. (BC.)

After having been largely silent for the first four days of proceedings, William Jennings Bryan (left, standing) rose to speak on Thursday, July 16. Bryan stated his support for Tennessee's antievolution law, criticized evolution because it undermines religion, denounced Darrow, and claimed that Scopes's student Howard Morgan understood the issues better than did Darrow. Seated to Bryan's left is Ben McKenzie, and in the right foreground is his son, Gordon McKenzie. Between the McKenzies is a WGN microphone. (UTK.)

On the fourth day of proceedings, Scopes was arraigned, after which chief counsel John Neal (standing with coat on) entered Scopes's plea: not guilty. Darrow is seated to the right of Neal, and Malone is seated to the left; John Scopes (in a white shirt and holding his head in his hand) is seated three people to the left of Neal. After entering Scopes's plea, the prosecution called its first witness: Rhea County school superintendent Walter White. The prosecution's case against Scopes lasted for less than two hours. (BC.)

The sixth day of the trial produced a heated exchange between Clarence Darrow (left) and Judge Raulston (right), which led to Raulston finding Darrow in contempt of court. When court reconvened, Darrow apologized, after which Raulston gave a religious speech in which he forgave Darrow. Raulston, who believed that the teaching of evolution would produce "mental stupidity and spiritual degeneracy," supported the Butler Act throughout his life. (BC.)

In this scene from court proceedings, Judge Raulston (left, with back to camera) addresses prosecutor Thomas Stewart (near center facing camera, with coat open) and defender Dudley Malone (by Stewart, with arms folded). Over Stewart's right shoulder are, from left to right, prosecutors Herbert Hicks and William Jennings Bryan, and to the left of Malone is Clarence Darrow (partly obscured). (BC.)

William Jennings Bryan (left) said relatively little during the first few days of proceedings, but he would not remain silent for long. Here, Bryan and Judge John Raulston chat one morning before court convened. (BC.)

In this scene from the first week of the trial, William Jennings Bryan (center left) leans over the table to talk with Clarence Darrow. Note that the jurors' chairs (right center) are empty, as they were for most of Scopes's famous trial. (UTK.)

During the Scopes Trial, Tennessee experienced one of its hottest summers on record. In this photograph, taken during a break on the trial's fourth day of proceedings, William Jennings Bryan cools off with a drink of water in the courtroom. During the trial's noon recess, WGN's Quin Ryan often aired interviews and "special speeches" by Bryan, George Rappleyea, John Scopes, F.E. Robinson, Judge John Raulston, and other people associated with the trial. (BC.)

Defendant John Scopes's contract at Rhea Central High School ended on May 1, 1925, but he stayed in Dayton because two of his students had been in a car accident, as well as to pursue "a beautiful blonde [he] had somehow previously overlooked." In this photograph, Scopes poses by one of the many "Read Your Bible" signs that were posted by evangelist T.T. Martin throughout Dayton during the trial. (UTK.)

During Scopes's trial, the trial's participants were often seen throughout town. In the photograph at right, defense attorney Arthur Garfield Hays (right, in white jacket and pants) chats with prosecuting attorney Herbert Hicks while defense lawyer Clarence Darrow, looking down and wearing suspenders, walks behind them. The Scopes Trial was Hays's first case with Darrow. (BC.)

On July 20, Judge Raulston held Clarence Darrow in contempt of court for Darrow's comments the previous week suggesting that Raulston favored the prosecution. When Darrow (far right, in suspenders) apologized, Raulston forgave Darrow, and they shook hands. Other attorneys shown here include Bryan (center, in white shirt with bow tie), Tom Stewart (partially obscured and to Bryan's right), Dudley Malone (behind WGN microphone), and Wallace Haggard (behind Malone with bow tie and cigar). (BC.)

Defense attorneys, expert witnesses, and journalists met most evenings at the "Mansion" to hear informal seminars and discuss the trial and strategy. This photograph shows, from left to right, an unidentified man, Frank Thone of the Science Service, biologist William A. Kepner from the University of Virginia, George Rappleyea, expert witness Tennessee geologist Wilbur Nelson, Watson Davis of the Science Service (barely visible), Charles Francis Potter, and Rabbi Herman Rosenwasser from San Francisco. The group on the right includes, from left to right, Arthur Garfield Hays, John Neal, John Scopes, and Dudley Field Malone. The chart between Hays and Rosenwasser is a scroll of a Hebrew translation of Genesis. Rosenwasser testified (via a statement read into the record by Hays) that other versions of the Bible—if translated accurately from the original Hebrew text—support evolution. Rosenwasser's testimony surprised many people at the trial, who did not know that the King James Version is not the only version of the Bible. (BC.)

Five

THE DEFENSE
CALLS MR. BRYAN

Hell is going to pop now.

—Dudley Field Malone to John Scopes,
as the defense called William Jennings Bryan to the witness stand, July 20, 1925

On Friday, July 17, after court was again opened with a prayer, Judge Raulston excluded the testimonies of expert witnesses—most of whom were scientists—that Darrow had brought to Dayton for the trial. Among the people upset with Raulston's decision was John Butler, who complained that Raulston "ought to give 'em a chance to tell [the court] what evolution is . . . I believe in being fair and square and American . . . Besides, I'd like to know what evolution is myself." When the defense objected to Raulston's decision, Raulston agreed to let the defense read the experts' statements into the record (for possible use during an appeal) but not in the presence of the jury. The infuriated Darrow, who, noting that Raulston had ruled for the state on virtually all of the trial's major issues, angrily suggested that Raulston was biased. At midmorning, Raulston recessed the trial to give the defense team's experts the time necessary to prepare their statements. July 17 was the shortest day of the trial.

When Raulston excluded Darrow's experts, most people felt the trial was all but over. H.L. Mencken—the greatest journalist of his era—told his readers, "Darrow has lost." Faced with enduring a hot weekend and hearing hours of testimony being read into the court record, Mencken and most other reporters left Dayton (in Scopes's words, "like birds in a migration"), believing there was nothing else to see. But when the Sunday edition of the *Nashville Banner* reported that the defense was "preparing to spring a coup d'état," some people began to wonder what might come next.

On Monday, July 20, Judge Raulston cited Darrow for contempt of court for comments Darrow had made the previous Friday. Darrow apologized, and Raulston—quoting scripture—forgave Darrow. For the rest of the morning and into the early afternoon, Arthur Garfield Hays—whose name was a mix of the names of three US presidents (with a different spelling for Hayes)—read experts' testimonies into the court record. Judge Raulston, concerned about the oppressive heat and the courtroom's sagging floor, reconvened the court on a dais under shade trees on the courthouse lawn. There, a crowd of almost 2,000 people witnessed the Scopes Trial's climactic event: Bryan being called by the defense to the witness stand. After sparring at a distance for more than a week, Darrow and Bryan were set to go head-to-head in a clash of faith and science, an event the *New York Times* described as "the most amazing courtroom scene in Anglo-American history."

INDEX No.

EDISON
RECORD

A PRODUCT OF
THE EDISON
LABORATORIES

51609-R
THE JOHN T. SCOPES TRIAL
**(THE OLD RELIGION'S BETTER
AFTER ALL)**
(Carlos B. McAfee)
Singing, Violin and Guitar
**VERNON DALHART AND
COMPANY**

10555

On July 10, 1925, popular country singer Vernon Dalhart (born Marion Slaughter) recorded "The John T. Scopes Trial (The Old Religion's Better After All)." A few weeks later, Dalhart also recorded "Bryan's Last Fight," which proclaims that William Jennings Bryan "stood for his own convictions, and for them he'd always fight." On the same day that Dalhart made his popular recording, the Scopes Trial opened with a prayer by fundamentalist preacher Lemuel M. Cartright that implored people to "be loyal to God." Dalhart's version of "The John T. Scopes Trial" was released four days after the Scopes Trial and sold more than 80,000 copies. (Randy Moore.)

The Scopes Trial generated countless trial-related products, including songs such as "The John T. Scopes Trial" and "You Talk Like a Monkey and You Walk Like a Monkey," which is shown here. In turn, the trial was inspired by the issue of the supposed evolutionary link between man and ape. This poster, which was printed a few years before the Scopes Trial, shows "Prof." Harry De Rosa and the monkey "Count the First," who is advertised as "Living Proof of the Darwin Theory." Neither De Rosa nor his trained monkey were in Dayton, but other monkeys and apes were in town for the big event, spurring the sale of many monkey-related souvenirs.

Several members of the defense team stayed at a dilapidated, 10-bedroom Victorian mansion just outside of Dayton. The house, which had been built in 1884 for executives and visiting stockholders of DCIC, had been vacant for most of a decade; it was referred to as the "Mansion" and the "Haunted House" by local residents. The defense team gathered at the Mansion each evening to plan strategy. The Mansion burned down in 1945. (BC.)

Scopes's defense team invited several scientists to Dayton to testify at Scopes's trial. This photograph shows, from left to right, (kneeling) Winterton Curtis, from the University of Missouri; Wilbur Nelson, state geologist of Tennessee; and William Goldsmith, from Southwestern University; (standing) Horatio Newman, from the University of Chicago; Maynard Metcalf, of John Hopkins University; Fay-Cooper Cole, from the University of Chicago; and Jacob Lipman, of the New Jersey Agricultural Station. (BC.)

To accommodate the expert witnesses, George Rappleyea and his wife, Ova Corvin "Precious" Rappleyea, renovated the abandoned "Mansion" across town. In this photograph, Precious stands on the steps of the renovated Mansion. Precious was a nurse who met George while he was recovering from a snakebite. The Rappleyeas, who married in 1919, moved to Dayton in 1922 and lived on a farm just outside of town. (SI Image No. 2008-1129.)

Luther Burbank, who was well known to Tennessee farmers, was a scientist famous for breeding new commercial varieties of plants. (This was considered to be evolution by some people.) Bryan worried that Burbank would testify for the defense, but Burbank did not come to Dayton. Burbank described the trial as "a great joke, but one which will educate the public and thus reduce the number of bigots." (LC.)

Maynard Mayo Metcalf, a 57-year-old zoologist from John Hopkins University, was the first and only witness called by the defense to testify on the stand at John Scopes's trial. The religious Metcalf, who testified on the afternoon of Wednesday, July 15, was meant to support the defense team's claim that science and religion can coexist. (In response, prosecutor Ben McKenzie claimed, "They want to put words in God's mouth.") The fourth day of Scopes's trial ended with Metcalf claiming, "The fact of evolution is as fully established as the fact that the earth revolves around the sun" and that "there is not a single [scientist] who has the least doubt of the fact of evolution." All of the other scientists that the defense team brought to Dayton submitted written statements that were read into the court record for the possible benefit of appellate courts. Scopes later claimed that the scientists in Dayton "broadened my view of the world." (LC.)

New York attorney Arthur Garfield Hays, pictured at right, was the newly appointed general counsel for the ACLU. On Monday, July 20, Hays—shown in the photograph below standing on the outside platform in a white shirt just to the left of the word "Bible"—read the experts' testimonies into the court record. Defense attorney Dudley Field Malone is seated behind Hays. Hays then announced the most dramatic event of the Scopes Trial: "The defense desires to call Mr. Bryan as a witness." After the trial, Hays—the only person at the ACLU who wanted John Scopes to be defended by Clarence Darrow—offered to send Judge Raulston a copy of Charles Darwin's *On the Origin of Species*, and Raulston agreed to accept it. (Both, BC.)

The climactic event in the Scopes Trial occurred on Monday, July 20, when Clarence Darrow—the most famous defense lawyer in the United States—questioned William Jennings Bryan for 90 minutes on a platform on the courthouse lawn. According to one report, Bryan often answered Darrow's questions by turning to the crowd, not to the jury, as he "tried to squeeze every drop of drama out of the part." When Bryan claimed that the questions by Darrow—whom Bryan called "the greatest atheist or agnostic in the United States"—were meant to ridicule people who believed in the Bible, Darrow snarled that he was merely trying to prevent "bigots and ignoramuses from controlling the education of the United States, and you know it." (Above, BC; below, LC.)

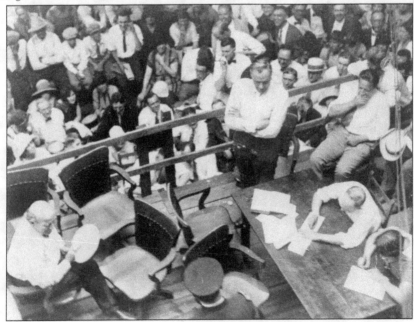

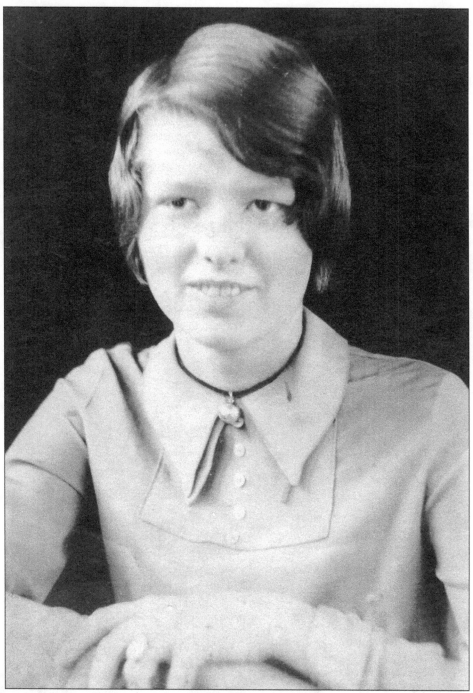

Eloise Purser (Reed), who turned 13 during the Scopes Trial, watched from the front row on July 20 as Darrow questioned Bryan on the courthouse lawn. Scopes, who was close friends with Eloise's brother Crawford (who years later described the Scopes Trial as "a wild time"), often ate dinner at the Purser home. Eloise was one of the last surviving attendees of Scopes's famous trial. In her later years, Eloise appeared in several documentaries about the Scopes Trial. (Donna Reed Taylor.)

A CIVIC BIOLOGY

Presented in Problems

BY

GEORGE WILLIAM HUNTER, Ph.D.

HEAD OF THE DEPARTMENT OF BIOLOGY, DE WITT CLINTON
HIGH SCHOOL, CITY OF NEW YORK.
AUTHOR OF "ELEMENTS OF BIOLOGY," "ESSENTIALS OF
BIOLOGY," ETC.

AMERICAN BOOK COMPANY
NEW YORK   CINCINNATI   CHICAGO

The biology class in which John Scopes was the substitute teacher used a textbook titled *A Civic Biology, Presented in Problems* by George William Hunter. This book was introduced as Exhibit 1 on the fourth day of court proceedings. On the following day, the pages of Hunter's book that received the most attention included page 194, which discusses evolution and depicts an evolutionary tree of animals. William Jennings Bryan cited that page, and F.E. Robinson—who owned the drugstore where the Scopes Trial conspirators met—testified that Scopes had admitted that he had taught the evolutionary tree shown on this page. Hunter's book, which claims that Caucasians were "the highest type of all" humans, had been officially adopted for use in Tennessee in 1919. The biggest impact of the antievolution movement was not on court decisions or legislative acts, but instead on biology textbooks. Indeed, after the Scopes Trial, the word *evolution* disappeared from biology textbooks in the United States. (William McComas.)

A *Civic Biology*, the first textbook to integrate zoology and biology, did not violate the Butler Act. Although Hunter classifies "man" as a mammal, he does not explicitly claim that classification implies common descent. "Man" does not appear in Hunter's evolutionary tree (right), nor are humans explicitly claimed to have any nonhuman ancestor. Instead, Hunter claims that modern humans evolved from "men who were much lower in their mental organization than the present inhabitants." (William McComas.)

As John Scopes's trial approached, geologist Leonard Darwin (right)—who would be the last surviving child of Charles and Emma Darwin—sent a letter to Scopes congratulating him for "his courageous effort to maintain the right to teach well-established theories . . . To state that which is true cannot be irreligious . . . May the son of Charles Darwin send you in his own name one word of warm encouragement." Leonard Darwin is shown here with paleontologist Henry Fairfield Osborn, who had several public skirmishes with William Jennings Bryan about evolution. Osborn was invited to Dayton to help the defense but could not come because of his wife's poor health. (LC.)

During the trial's second weekend, most reporters (including H.L. Mencken) left Dayton because they believed the trial was virtually over. They were wrong. Those who left early missed Darrow (right) in his legendary questioning of Bryan (left) on the platform outside the courthouse, which was watched by almost 2,000 spectators. Bryan, who believed that children lose interest in the Bible "when they come under the influence of a teacher who accepts Darwin's guess," proudly defended the Bible during Darrow's questioning about Noah and the flood, "the big fish [that] swallowed Jonah," Joshua commanding the sun to stand still, human ancestors, and the age of Earth. After the questioning, John Scopes noted that "Bryan was never the same." Scopes later praised Darrow's work, noting that "if men like Clarence Darrow had not come to my aid and had not dramatized the case to a responsive world, freedom would have been lost." Scopes was conscripted to write reports for the absent reporters. (LC.)

Six

THE VERDICT, BRYAN'S DEATH, AND THE APPEAL

We see nothing to be gained by prolonging the life of this bizarre case.

—Tennessee Supreme Court, when overturning John Scopes's conviction

Near the end of Darrow's relentless and often contentious questioning, Bryan—clearly playing to the crowd—proclaimed that he was in Dayton to defend religion and that Darrow's only goal was "to slur at the Bible." Darrow angrily responded, telling Bryan, "I am [examining] you on your fool ideas that no intelligent Christian on Earth believes . . . You insult every man of science and learning in the world because he does not believe in your fool religion."

Bryan had fallen into the trap that Darrow tried to spring two years earlier with his public letter in the *Chicago Tribune*. Thomas Stewart, realizing Bryan's no-win predicament and wondering how what he was seeing could be relevant to Scopes's innocence or guilt, tried repeatedly to stop Darrow's questioning, but Bryan—having a larger stake in the questioning than anyone—knew he could not step away. When Bryan then accused Darrow—whom Bryan labeled "the greatest atheist or agnostic in the United States"—of ridiculing people who believe in the Bible, Darrow responded that he was merely trying to prevent "bigots and ignoramuses from controlling the education of the United States, and you know it." After fending off Darrow's questions about a worldwide flood, Jonah being swallowed by a whale, and Joshua commanding the sun to stand still, Bryan then did what many fundamentalists considered unfathomable—when asked about Earth's age, Bryan admitted that Earth was not made in "six days of twenty-four hours . . . it might have continued for millions of years." In doing so, Bryan admitted that he, too, interpreted (rather than took literally) parts of the Bible and, by implication, that perhaps evolution too could be part of such an interpretation. Many in attendance were shocked by Bryan's answers, which Scopes described as "astonishing. This was the great shock that Darrow had been laboring for all afternoon." Judge Raulston then abruptly ended the day's proceedings.

The following morning, Judge Raulston struck Bryan's testimony from the court record because he felt it would be irrelevant during an appeal. Darrow then abruptly asked Judge Raulston "to bring in the jury and instruct [them] to find the defendant guilty." Because Darrow did not give a closing argument, Bryan was not allowed to deliver his long-anticipated closing speech either. Bryan agreed to be questioned by Darrow on the condition that he would be allowed to question Darrow, but the trial's abrupt end denied this to Bryan also. Judge Raulston then charged the jury, and they reconvened outside to decide John Scopes's fate.

After the jury announced Scopes's guilt, Judge Raulston fined Scopes $100. Then Scopes addressed the court for the first and only time: "I feel that I have been convicted of violating an unjust statute. I will continue in the future, as I have in the past, to oppose this law in any way I can. Any other action would be in violation of my ideal of academic freedom—that is to teach the truth as guaranteed in our Constitution of personal and religious freedom. I think the fine is unjust." In this photograph taken at sentencing, Scopes is surrounded by Rhea County sheriff Robert "Bluch" Harris (left) and Deputy Kelso Rice (right), a police officer from Chattanooga that Judge Raulston appointed bailiff during Scopes's trial. (Both, BC.)

The Dayton Herald

DAYTON, TENNESSEE, THURSDAY, JULY 23, 1925 — NUMBER 7

OSION ENTOMBS
MEN AT ROCKWOOD

...trs known the fatal time come and little hope
...e Iron Co., is maintained for the lives of the en-
...s morning tombed men. When those who are
...acquainted with the mines were ask-
...ed their opinion of the safety of the
... mine en- men they shook their heads in grave
...st citizens. fear.
...in the mine
...y had been State inspectors and res'ue part-
...ut, headed ies from all this section has been
...he morning summoned.

ISPECT — FIRE DESTROYS BOX CARS

Mrs. W. F. Fire which was supposed to have

TENNESSEE WEEKLY INDUSTRIAL REVIEW

Martin—$20,000 Sunday school annex to Methodist church finished.

Crossville—Tennessee Electric Power Co., makig rapid progress toward installing power lines here.

Martin—American Legion rents newly remodeled quarters over Garretts & Walker's store for clubrooms erectingAD a ffff. fei etaoln shdi g

Covington—$25,000 bond issue voted in connection with addition to Byars-Hall High School.

Lavergne—Plans under way for erecting new school here.

Memphis—Library of Goodwyn Institute to be repaired after fire and many books will be rebound.

Keefe—New Christian Church finished.

PROMINENT CITIZEN DIES AFTER LINGERING ILLNESS

William F. Blevins died at the home of his daughter, Mrs. J. R. Gillespie, on Main st., Wednesday morning, July 15.

He served as first lieutenant of Company I, Fifth Tennessee cavalry during the civil war. He was a charter member of the J. W. Gillespie camp of Confederate Veterans. He was appointed clerk and master of Meigs county by Judge Key and held that position for twelve years, after which he engaged in the mercantile business for many years.

He was born near Decatur, Tenn., Oct. 27, 1835. He married Miss Mary Russell, of Decatur, who died in 1920. He is survived by three

JURY RETURNS VERDICT OF
"GUILTY" IN SCOPES CASE

The curtain dropped Tuesday noon on the world-watched, nation-wide drama entitled "The State of Tennessee vs. John Thomas Scopes," when the jury brought in a verdict of guilty and recommended a fine of $100. It was rather anti-climax; —a relaxed finish to the bitterest legal battle ever waged in the United States—, a foretold outcome to a conflict that, earlier in the proceedings, had keyed to their highest pitch the keenest, shrewdest minds of the

worlds greatest advocates of evolution. The bitterness of Darrow over the failure to get these expert witnesses on the stand exploded itself in his being cited for contempt—which citation was next day withdrawn when Darrow, after apologies to the court.

On account of the vast crowds here to hear the trial, Monday's session was held on a dias on the lawn;

After the verdict, Bryan attributed the trial's popularity to the fact that "causes stir the world" and that the trial would bring issues discussed in Dayton "before the attention of the world." Darrow, however, described the trial as a fight against "testing every fact in science by a religious dictum." When Judge Raulston then adjourned the court, people in the courtroom began milling around while discussing Scopes's famous case. In the above photograph, Sheriff Harris (standing left of the WGN microphone) watches as John Scopes speaks with his attorney Dudley Malone (right). On the other side of the courtroom (below), prosecutor Ben McKenzie (center left, in white jacket with his arm around Darrow)—who called all of the defense attorneys from the north "foreigners"—visits with Clarence Darrow, with whom he had became close friends. When McKenzie was later charged with transporting bootleg whiskey, Darrow offered his services, but McKenzie declined. (Both, BC.)

On the weekend after the trial, Bryan went to Winchester, Tennessee, where he gave a 90-minute speech to 6,000 people (the largest crowd for an event in Winchester's history). That speech, which would be Bryan's last, ended with the words, "Faith of our fathers—holy faith, we will be true to thee 'til death." Bryan then went to Chattanooga to talk with *Chattanooga News* editor George Fort Milton about printing the speech that he was not allowed to deliver at the trial. (In that speech, which was meant to rival his "Cross of Gold" speech, Bryan claims that evolution contradicts the Bible, destroys faith, diverts attention from spiritual matters, and damages society.) After spending the night of July 25 in Chattanooga's newly opened Hotel Ross, Bryan was driven by Herschel Keith to Dayton, where he appeared unannounced on Sunday morning at First Southern Methodist Church. Bryan led a prayer but did not preach. This photograph, taken that morning by Keith, is the last photograph of Bryan. (BC.)

After attending church in Dayton, Bryan returned to the Rogers' home, ate a large lunch, and lay down for a nap. That afternoon, Mary Bryan sent William Hugh "Jimmy" McCartney (the family's chauffeur) to awaken her husband so Bryan could prepare for his sermon that evening at First Southern Methodist Church. McCartney could not awaken Bryan. The photograph at right is of Walter Fairfield "Doc" Thomison, who lived just down the street from the Rogers' home; Thomison pronounced Bryan dead from apoplexy at 3:30 p.m. on July 26, 1925. After Bryan's death was announced, crowds gathered at the Rogers' home (below); George Rappleyea was one of the first people to visit. Bryan's body was prepared for burial by Dayton's Coulter Funeral Home on Market Street. Bryan's memorial service in Dayton was held on the lawn of the Rogers' home on the afternoon of July 28. The service was officiated by Rev. Charles R. Jones (pastor of Dayton's First Southern Methodist Church), who had given the benediction at the Scopes Trial. (Right, Pat Hawkins Guffey; below, BC.)

In the above photograph, Bryan's casket arrives at the home of F.R. Rogers, where Bryan and his entourage had stayed for more than two weeks. Although crowds in Dayton dispersed soon after Scopes's trial, the area continued to mourn Bryan's death. The photograph below, with a view looking south down Market Street in Dayton, was taken on July 30, 1925, the day that Bryan was buried at Arlington National Cemetery in Virginia. Pres. Calvin Coolidge ordered that all American flags be flown at half-staff that day in tribute to Bryan, who had once been secretary of state. In the aftermath of his death, Bryan became a martyr compared to Christ, while Scopes's defenders were likened to King Herod, Pontius Pilate, and other biblical villains. (Both, BC.)

After ceremonies in Dayton, Bryan's flag-draped casket was taken to the train station, where a crowd had gathered, as seen above. In the photograph below, an honor guard assembled by Dayton's American Legion post loads Bryan aboard a special Pullman car for transport (with stops in Knoxville, Bristol, Roanoke, Lynchburg, and other cities) to additional ceremonies in Washington, DC. The short man (in uniform and wearing riding boots) in the center of the group helping raise the coffin is trial instigator George Rappleyea, who had served in World War I and was post commander of American Legion in Dayton in 1925. (Both, BC.)

On July 30, Bryan's body arrived in Washington, DC, where it lay in repose at the New York Avenue Presbyterian Church, often called "The Church of the Presidents." Bryan's funeral in Washington, DC, which was officiated by Rev. Joseph R. Sizoo, was held on July 31 and was broadcast nationwide on radio. Mourners waited in line for hours to pay their respects to Bryan and see his glass-topped casket at the church. Bryan was buried during a rainstorm atop a hill in Arlington National Cemetery. (Above, BC; below, Randy Moore.)

While he was in Dayton, William Jennings Bryan mentioned that he wanted to establish a college in Dayton "to teach truth from a Biblical perspective." The next year, Tennessee governor Austin Peay—who signed the Butler Act into law—broke ground on that school atop a hill overlooking Dayton. Peay, like Bryan, opposed the teaching of evolution because he believed that it poisoned young people's minds. (UTK.)

The first student to enroll in William Jennings Bryan University was Amy Cartright (Robinson), who graduated from Rhea Central High School in 1930. Cartright's grandfather, Methodist preacher Lemuel M. Cartright, opened the first day of the Scopes Trial with a fundamentalist prayer that John Scopes described as "interminable." Later that day, after seeing the spectacle surrounding his trial, Scopes simply commented, "This is incredible." (BC.)

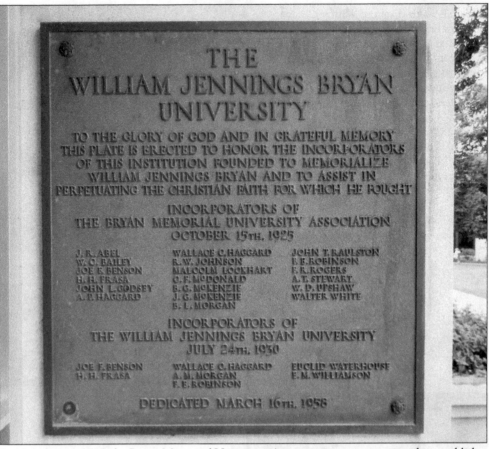

On October 15, 1925, the Bryan Memorial University Association was incorporated to establish a college in Dayton in Bryan's honor. As noted on this plaque later mounted at that college (today's Bryan College), the incorporators included many people associated with the Scopes Trial, including Wallace Haggard, Ben McKenzie, John Raulston, F.E. Robinson, F.R. Rogers, Thomas Stewart, and Walter White. When William Jennings Bryan University opened in 1930, its first classes met in the abandoned Rhea Central High School building where John Scopes allegedly taught evolution just five years earlier. In October 1935, the first classes were held at Bryan College's current location in what became Mercer Hall, a building with the same dimensions as Noah's ark. Today, Bryan College is a nondenominational, liberal arts college that endorses the infallibility of the Bible and rejects human evolution. (Above, Randy Moore; below, William McComas.)

Many organizations and individuals tried to capitalize on Bryan's death. One such group was the Ku Klux Klan. Several leaders of the fundamentalists' antievolution crusade (e.g., Frank Norris and Billy Sunday) were sympathetic to the Ku Klux Klan (KKK), and some critics attributed the support for Butler-like laws in the South in the 1920s to a fear of racial equality between whites and blacks. Bryan was never a KKK member, and he rejected many of the KKK's views. However, at the 1924 Democratic National Convention, Bryan had argued strongly against a resolution condemning the Klan. Consequently, when Bryan died, the KKK burned crosses in Bryan's memory, and at a rally in Washington, DC, two weeks after his death, Bryan was eulogized as "the greatest Klansman of our time." (Both, LC.)

At one point, Scopes had hoped to become an attorney, but his son John Scopes Jr. later admitted that his famous father "thought everyone would compare him to Clarence Darrow, and that was too much pressure for him." Thus, after his famous trial, Scopes moved to Chicago, where he used $2,550 raised by scientists and reporters at his trial to study geology for two years at the University of Chicago. Soon thereafter, Scopes accepted a job with Gulf Oil and began working in the oil fields 70 miles east of Maracaibo, Venezuela. This photograph shows Scopes in 1927 in Venezuela. In a letter written from there on November 15, 1927, to Clarence and Ruby Darrow, Scopes confesses, "I hate office work and city life." On his way to Venezuela, Scopes stopped in New York, where he saw "several people that helped put Dayton on the map." (Riesenfeld Rare Books Research Center, University of Minnesota Law Library.)

Seven

A REUNION IN DAYTON

The evolution controversy will go on, with other actors and other plays.

–John Scopes, after seeing *Inherit the Wind*

When Scopes's trial ended, Rhea Central High School offered to rehire Scopes if he would "adhere to the spirit of the evolution law." Scopes, who considered his year in Dayton "one of the most pleasant of my life," declined the offer and used a scholarship created by his trial's reporters and expert witnesses to enroll in graduate school at the University of Chicago. (Scopes's successor in Dayton was Raleigh Valentine Reece, a fundamentalist and newspaper reporter from Nashville.) Scopes then worked as a petroleum engineer for Gulf Oil in Venezuela for three years, after which he returned to the United States and lost a run for Congress in Kentucky (as a candidate of the Socialist Party). Scopes then resumed his career in the oil industry. Scopes, wanting to be "just another man instead of the Monkey Trial defendant," avoided reporters ("I am tired of fooling with them"), shunned publicity, and seldom spoke about his accidental fame. It was the ACLU, not Scopes, who paid the $343.80 in court costs; the $100 fine levied by Judge Raulston at the end of the trial was never collected.

As both sides claimed victory in Dayton, two more states—Mississippi and Arkansas—passed laws banning the teaching of human evolution. John Butler, who had written Tennessee's antievolution law, proclaimed the Scopes Trial "the beginning of a great battle . . . the controversy of the age," and organizations such as the Supreme Kingdom were formed "to carry to every nook and corner of the world the fight against the Darwinian theory." But elsewhere, the country was growing tired of the evolution controversy; antievolution bills failed in many states, and even William Bell Riley—the preacher who organized and led the antievolution crusade of the 1920s—could not garner enough support to pass a Butler-like bill in his home state of Minnesota. Meanwhile, as a result of the Scopes Trial, the word *evolution* disappeared from biology texts and classrooms, even though the antievolution laws' focus—that is, *human* evolution—is but an example of the most important principle in biology. Although the Scopes Trial is often portrayed as the pivotal event in the evolution-creationism controversy, it brought about a major decline in the teaching of evolution whose recovery did not begin until the early 1960s.

In decades following the Scopes Trial, the trial's legend began to grow. In 1950, the play *Inherit the Wind* used a fictitious retelling of the Scopes Trial to expose the demagogic challenge to intellectual freedom presented by McCarthyism. (The play's title quotes Proverbs 11:29: "He that troubleth his own house shall inherit the wind.") The movie version of the play was shown at the Dayton Drive-In Theater on July 21, 1960, as part of Scopes Trial Day, Dayton's second attempt to cash in on Scopes's famous trial. Interestingly, the film was the first-ever in-flight movie shown on some TWA flights not long after it premiered. The years had softened many people's concerns about the teaching of evolution, and Scopes—at the urging of his wife—returned to Dayton a hero.

On July 21, 1960, John Scopes—then 59 years old—returned for the first time in more than three decades to the town that made him famous. Scopes came to Dayton for Scopes Trial Day, which coincided with the 35th anniversary of his conviction. While Scopes was in Dayton, a local preacher denounced him as "the Devil," and Scopes noted that teachers were required to pledge that they would not teach evolution. Scopes remained skeptical of government intrusion into the classroom, noting that "it is as bad to have school and state joined together as it is to have church and state joined together." Scopes Trial Day produced the second-largest crowd in Dayton's history. (BC.)

During Scopes Trial Day festivities, Scopes is given a key to the city by Dayton mayor J.J. Rodgers (above, left). Between Rodgers and Scopes is Scopes's wife, Mildred, whom he met while working in Venezuela and married in 1930. (As part of his commitment to the marriage, Scopes was baptized in the Roman Catholic Church of Venezuela; this was the only church with which Scopes ever had any formal relationship.) During Scopes Trial Day events, John and Mildred Scopes (below) enjoy a drink at Robinson's Drug Store, where they sit at the small table at which Scopes's famous trial originated. Shown here with the Scopeses is J.M. "Mack" Jones, a local businessman (right). (Both, BC.)

During the Scopes Trial Day festivities, John Scopes was greeted by Miss Dayton, Nancy Shipley. During the celebration, Scopes told reporters that if he could replay his time in Dayton in 1925, he "would do [the trial] again," but admitted that he "had hoped the trial would accomplish more than it did." As Scopes had earlier lamented, "I can't understand why, in this Space Age and in view of all the advancements science has achieved, we are still arguing [about] evolution." (Nancy Shipley.)

Dayton's Scopes Trial Day celebration in 1960 reunited several participants and others from Scopes's famous trial. Shown in the above photograph are, from left to right, prosecutor Sue Hicks, reporter Earl Shaub, John Scopes (leaning on original witness chair), prosecutor Wallace Haggard, and an unidentified reporter who covered the trial. In the photograph below, Scopes (right) chats with Bryan College's fourth president, Theodore C. "Ted" Mercer (left), and Helen Rudd Brown, William Jennings Bryan's granddaughter. (Above, UTK; below, BC.)

In this photograph, taken during Scopes Trial Day festivities, John Scopes examines a 1909 Ford that was part of the parade held in his honor. Behind Scopes is Robinson's Drug Store, which had moved from its original location on Main Street (where the Scopes Trial was instigated) to a newer building on Market Street near the Rhea County Courthouse. (SI.)

During Scopes Trial Day celebrations, John Scopes (right) visited the new Robinson's Drug Store. There, he was greeted by W.C. "Sonny" Robinson (left), the only son of F.E. and Clark (Haggard) Robinson. (Also shown is Sonny's son Bill.) Scopes and Robinson reminisced about Robinson's father, F.E., a Dayton civic leader who had advertised himself for decades as "The Hustling Druggist." Although the elder Robinson had died three years before the reunion, the drugstore remained a center of activity in Dayton. (BC.)

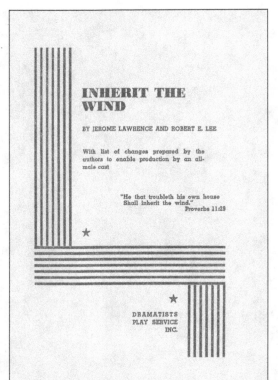

Inherit the Wind is a play and movie that uses a fictitious retelling of the Scopes Trial to warn of the dangers of religious and political zealotry. The play, which was written in 1950 by Jerome Lawrence and Robert E. Lee, premiered to favorable reviews on January 10, 1955, in Dallas, Texas, and then at Broadway's National Theatre on April 21, 1955, where its three-year run made it the longest-running drama on Broadway at that time. The playwrights of Inherit the Wind emphasize that "Inherit the Wind is not history" and "does not pretend to be journalism."

The movie version of Inherit the Wind, which was directed by Stanley Kramer, was released in 1960. Before filming, Kramer asked prosecutor Tom Stewart to serve as a consultant for the movie, but Stewart declined the offer. The award-winning movie garnered critical acclaim, but was a box-office failure. Since 1960, there have been several remakes of Inherit the Wind, but none has had the impact of the original version.

IN HONOR OF
SCOPES TRIAL DAY
YOU ARE CORDIALLY INVITED TO THE
FIRST PUBLIC U. S. PRESENTATION
OF
STANLEY KRAMER'S
"INHERIT THE WIND"
STARRING
SPENCER TRACY
FREDRIC MARCH
GENE KELLY
WITH
DICK YORK — DONNA ANDERSON
AND FLORENCE ELDRIDGE

PRODUCED AND DIRECTED BY STANLEY KRAMER

A UNITED ARTISTS RELEASE

DAYTON DRIVE-IN THEATRE
THURSDAY, JULY 21, 1960 — 8:30 P.M.

ADMISSION BY
RESERVATION ONLY
R.S.V.P.

J. J. RODGERS, M.D.
MAYOR
DAYTON, TENNESSEE

The Scopes Trial Day festivities included the American premiere of the movie version of *Inherit the Wind* (directed by Stanley Kramer) at the Dayton Drive-In Theater. Contrary to what was advertised in Dayton, this showing was actually the film's American premiere (it had been shown earlier at the Berlin Film Festival). The photograph above shows the invitation to see *Inherit the Wind* that was sent by the Dayton Chamber of Commerce to Scopes's prosecutor Sue Hicks. Hicks believed that *Inherit the Wind* was "a travesty on William Jennings Bryan" and had to be dissuaded by his family from buying television time to counter the movie's claims. After seeing *Inherit the Wind*, John Scopes promised that the evolution controversy would "go on, with other actors and other plays." He was right; the 127-minute-long movie had a far greater impact on the public's view of the Scopes Trial than did the trial itself, even if what people saw in the movie was often more fiction than fact. (Above, UTK; below, Dean Wilson.)

The movie version of *Inherit the Wind* captivated audiences worldwide. In the above photograph, the Darrow-like character (left, played by Spencer Tracy) questions the Bryan-like character (played by actor Fredric March) in a scene using dialogue taken almost verbatim from the court transcript. In the photograph below, the Bryan-like character dies in the courtroom as the trial concludes. Although this event, like many others in *Inherit the Wind*, did not happen at Scopes's trial, John Scopes believed that the movie "captured the emotions in the battle of words between Bryan and Darrow." (Both, United Artists.)

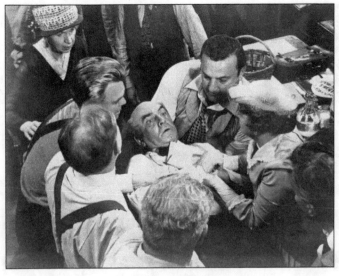

Eight

BEYOND THE SCOPES TRIAL

The cause defended at Dayton is a continuing one that has existed throughout
man's brief history and will continue as long as man is here. It is the
cause of freedom for which each man must do what he can.

—John T. Scopes

The demise of the Butler Act began, oddly enough, not in Tennessee, but in another southern state, Arkansas. There, a young biology teacher named Susan Epperson challenged her state's Scopes Trial–influenced ban on the teaching of human evolution, which had passed in 1928 in a public referendum. Epperson's case eventually reached the US Supreme Court, which ruled unanimously in late 1968 in *Epperson v. Arkansas* that the Arkansas law banning the teaching of human evolution was unconstitutional.

Amidst emotional professions of faith and countless references to the Scopes Trial, Tennessee repealed the Butler Act on May 18, 1967. For the first time in more than 40 years, it was again legal to teach human evolution in Tennessee's public schools. In late 1970, the nation's last remaining law banning the teaching of human evolution—that is, the one in Mississippi inspired by the Scopes Trial—was declared "void and of no effect" by the Mississippi Supreme Court. When Tennessee repealed its antievolution law, John Scopes—making a rare public appearance—said, "Better late than never," expressing his pleasure that the law used to prosecute him had been overturned.

Although there were no more attempts to ban the teaching of evolution, creationists soon returned to court with new strategies to force their ideas into biology classrooms of public schools. These attempts involved claims that evolution is offensive to religious sensibilities and therefore should not be taught (*Wright v. Houston Independent School District*, 1972), that taxpayers' money should not be used to fund textbooks or public displays that promote evolution (*Willoughby v. Stever*, 1973), that teachers must give "equal time" to "creation science," a form of creationism (now called "young-Earth creationism"; *McLean v. Arkansas Board of Education*, 1982; *Edwards v. Aguillard*, 1987), that teachers have a free-speech right to teach creationism (*Webster v. New Lenox School District*, 1990; *Peloza v. Capistrano Unified School District*, 1992), and that teachers can disguise creationism as "intelligent design" (*Kitzmiller v. Dover Area School District*, 2005). At one point or another, each of these lawsuits mentioned John Scopes's famous trial from decades earlier. Creationists lost all of these challenges.

The Scopes Trial and its legend continue to profoundly influence legislation, court decisions, and local skirmishes about the evolution-creationism controversy. The Scopes Monkey Trial remains the standard to which all other court cases involving evolution and creationism are compared.

After working in Venezuela, John Scopes—who wanted to be "just another man instead of the Monkey Trial defendant"—returned to the United States and worked for United Gas in Houston, Texas, and Shreveport, Louisiana. Scopes continued to decline opportunities to profit from his trial because "people wouldn't come to listen; they'd come to look, and I don't care to be an exhibit." This photograph shows ScopesScopes in Shreveport in 1950, at which time he told a reporter, "My friends who know about [the evolution trial] never bring it up in my presence." (Jerry Tompkins.)

On January 17, 1927, the Tennessee Supreme Court—consisting of, from left to right, (first row) William Cook, Grafton Green, and Alexander Chambliss; (second row) William Swiggart and Colin McKinney—ruled the Butler Act constitutional but set aside John Scopes's conviction on a technicality: Judge Raulston fined Scopes $100, but the Tennessee Constitution required all fines greater than $50 to be levied by a jury. As the court noted, "We see nothing to be gained by prolonging the life of this bizarre case." (TSLA.)

In 1967, Tennessee House Bill 48—a bill repealing the 42-year-old Butler Act—passed the House by a vote of 58-27, and it was then passed by the Senate by a vote of 20-13. On May 18, 1967, Tennessee governor Buford Ellington signed the legislation into law. John Scopes was the only teacher ever prosecuted under the Butler Act. This photograph shows Ellington soon after signing the repeal of the Butler Act. (TSLA.)

Arkansas Forecast
Fair and warmer Wednesday
with increasing cloudiness and a
chance of showers Thursday.
(Weather Map on Page 9B.)

azette.

, 1968.   38 Pages   * *   10 Cents

Arkansas Law On Evolution Struck Down

It Violates Amendments, Justices Say

By ED JOHNSON
Gazette Washington Bureau
314 National Press Building

WASHINGTON — The United States Supreme Court struck down Arkansas's antievolution law Tuesday on the ground that it tears down in the public school system the wall between church and state guaranteed by the United States Constitution.

The decision means that the public school teachers in the state are free to teach Darwin's theory that mankind evolved from another animal or animals, without fear of criminal prosecution.

The "monkey law" has been

Victor in the Evolution Case

Mrs. Susan Epperson is shown with her father, T. L. Smith, during the trial two years ago. Like his daughter he also is a teacher of biology.

Arkansas's law banning the teaching of human evolution was still on the books when, in 1964, Susan Epperson began teaching biology at Central High School in Little Rock, Arkansas. Epperson challenged the law because she believed it was "a sure path to the perpetuation of ignorance, prejudice, and bigotry," and her court brief ends with a dramatic reference to "the famous Scopes case" and "the darkness in that jurisdiction" that followed Scopes's conviction. In 1968, while citing "the celebrated Scopes case," the US Supreme Court ruled unanimously in *Epperson v. Arkansas* that banning the teaching of evolution is unconstitutional (left). As a result of *Epperson v. Arkansas*, all laws banning the teaching of human evolution in public schools were overturned by 1970. On January 6, 1969, John Scopes met Susan Epperson (below) in Shreveport, Louisiana. Scopes confided to her that he was "very happy about [the Supreme Court's] decision in her case," adding that he had "thought all along—ever since 1925—that the law was unconstitutional." (Both, Susan Epperson.)

When he retired in 1963, John Scopes was still receiving an average of one letter per day about his trial. Throughout his life, Scopes received lucrative offers to cash in on his fame, but—except for his book *Center of the Storm*—he rejected all offers because he "had too much respect for the issues involved in the trial." This photograph shows Scopes in his home in Shreveport, Louisiana, around 1965. (Jerry Tompkins.)

In 1967, John Scopes published (with James Presley) *Center of the Storm: Memoirs of John T. Scopes*. While promoting his book on television shows such as *The Today Show* and *The Tonight Show with Johnny Carson*, Scopes often commented that "the Bible simply isn't a textbook of science, and that's all there is to it." (Randy Moore.)

On April 1, 1970, John Scopes returned to a Tennessee classroom for the first time in 45 years when he spoke to an overflow crowd at the George Peabody College for Teachers (now a part of Vanderbilt University) about his famous trial. The self-deprecating Scopes, who was greeted "like a returning hero," admitted that he had done "little more than sit, proxy-like, in freedom's chair, that hot, unforgettable summer," something that he admitted was "no great feat, despite the notoriety it has brought me." (Vanderbilt University.)

During his visit to the George Peabody College for Teachers, Scopes reiterated his beliefs that education should be free of "the contamination of the state," that academic freedom is "a cause of freedom for which each man must do what he can . . . as long as man is here," and that "it is the teacher's business to decide what to teach." This was Scopes's last public appearance. (Steve Smartt.)

In the photograph at right, taken on July 21, 1925, the youthful John Scopes hears that he has been convicted of violating Tennessee's Butler Act. Scopes subsequently avoided publicity and seldom talked about his famous trial, yet he became a reluctant hero as his trial became a symbol of the importance of academic freedom. After leading a private life and fathering two sons, Scopes—a heavy smoker for most of his life—died in October 1970 of cancer in his home near Shreveport, Louisiana. He was buried in a family plot in Oak Grove Cemetery in Paducah, Kentucky, not far from Lone Oak School, where he first learned about evolution. Scopes rests beneath the inscription "A Man of Courage," a phrase used by Clarence Darrow to describe Scopes. Nearby rests Scopes's sister Lela, who was fired as a math teacher in Paducah when she refused to renounce her brother's beliefs. (Right, BC; below, Randy Moore.)

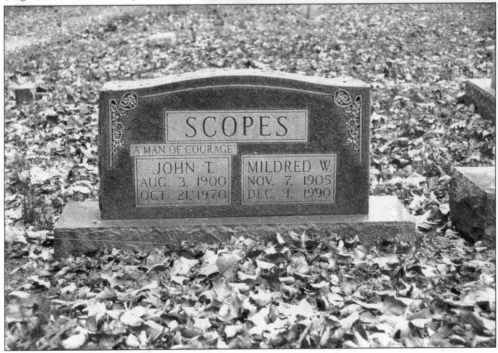

Every July, Dayton sponsors the Scopes Trial Play and Festival, which includes a reenactment of Scopes's famous trial. (Above, Randy Moore; below, Dean Wilson.)

Although the Scopes Monkey Trial took place in 1925, its legend continues to lure people to Dayton, Tennessee, where the trial occurred. This historical marker stands in front of the Rhea County Courthouse, the site of John Scopes's famous trial, welcoming visitors to the site of the "trial of the century." (Randy Moore.)

BILIOGRAPHY

Burgan, Michael. *Perspectives on the Scopes Trial: Faith, Science and American Education.* Tarrytown, NY: Marshall Cavendish, 2011.

de Camp, L. Sprague. *The Great Monkey Trial.* Garden City, NY: Doubleday, 1966.

Gaillard, Frye. "Revisiting Scopes." *Peabody Reflector* 75 (Fall 2006): 21–25.

Ginger, Ray. *Six Days or Forever: Tennessee v. John Thomas Scopes.* New York: New American Library, 1958.

Hunter, George William. *A Civic Biology: Presented in Problems.* New York: American Book Company, 1914.

Johnson, Anne Janette. *Defining Moments: The Scopes "Monkey Trial."* Detroit: Omnigraphics, 2007.

Kearper, Steve. "Evolution on Trial." *Smithsonian* 36 (2005): 52–61.

LaFollette, Marcel C. *Reframing Scopes: Journalists, Scientists, and Lost Photographs from the Trial of the Century.* Lawrence: University Press of Kansas, 2008.

Larson, Edward J. *Summer for the Gods: The Scopes Trial and America's Continuing Debate over Science and Religion.* Cambridge, MA: Harvard University Press, 1997.

Lawrence, Jerome, and Robert E. Lee. *Inherit the Wind.* New York: Bantam Books, 1955.

Moore, Randy. *In the Light of Evolution: Science Education on Trial.* Reston, VA: National Association of Biology Teachers, 2000.

Robinson, F.E., and W.E. Morgan. *Why Dayton, of All Places?* Chattanooga: Andrews Printery, 1925.

Scopes, John T., and James Presley. *Center of the Storm: Memoirs of John T. Scopes.* New York: Holt, Rinehart and Winston, 1967.

Tompkins, Jerry R., ed. *D-Days at Dayton: Reflections on the Scopes Trial.* Baton Rouge: Louisiana State University Press, 1965.

The World's Most Famous Court Trial: Tennessee Evolution Case. Dayton, TN: Bryan College, 1990.

INDEX

Visit us at
arcadiapublishing.com